Engineering Circuit Analysis

Engineering Circuit Analysis

Jasper Harrison

NY RESEARCH
PRESS

New York

Published by NY Research Press
118-35 Queens Blvd., Suite 400,
Forest Hills, NY 11375, USA
www.nyresearchpress.com

Engineering Circuit Analysis
Jasper Harrison

International Standard Book Number: 978-1-63238-695-3 (Hardback)

Cataloging-in-Publication Data

Engineering circuit analysis / Jasper Harrison.
    p. cm.
Includes bibliographical references and index.
ISBN 978-1-63238-695-3
1. Electric circuit analysis. 2. Electric network analysis. I. Harrison, Jasper.
TK454 .E54 2019
621.319 2--dc23

Contents

Preface

An electronic circuit is a framework of electronic components like capacitors, resistors, transistors, diodes, etc. that are connected by wires through which an electric current can flow. It can be an analog circuit, a digital circuit or a mixed-signal circuit. Analog circuits are those in which current or voltage varies continuously with time. Some of the basic components of analog circuits are resistors, capacitors, inductors, wires, etc. Analog circuit analysis uses Kirchhoff's circuit laws. In digital circuits, electric signals have discrete values. Transistors are interconnected to create logic gates that provide the functions of Boolean logic. Mixed-signal circuits consist of elements of both analog and digital circuits. Examples are analog-to-digital converters, digital-to-analog converters, etc. Network analysis refers to the process of determining the currents and voltages across every component in a network. Network analysis can be done using the methods of nodal analysis, mesh analysis, superposition and effective medium approximations. This book is a valuable compilation of topics, ranging from the basic to the most complex theories and principles in the field of engineering circuit analysis. Most of the topics introduced herein cover new techniques of circuit analysis and their applications in a comprehensive manner. For all those who are interested in this field, this book can prove to be an essential guide.

A short introduction to every chapter is written below to provide an overview of the content of the book:

Chapter 1 - A circuit is an electrical network with a closed loop that provides a return path for the current. The aim of this chapter is to explore the different aspects of circuits and their different types, such as electronic and electrical circuits, and series and parallel circuits;

Chapter 2 - Any discrete electrical device that is used in an electronic system or circuit to affect the motion of electrons and their associated fields is termed as an electrical component. Some common circuit components are capacitors, resistors, inductors, voltage sources, switches, etc. which have been extensively covered in this chapter;

Chapter 3 - To develop a comprehension of circuit designing, it is vital to understand the fundamental electrical laws and theorems for circuit designing. This chapter includes various topics central to the understanding of these aspects such as Kirchhoff's current law, Ohm's law, Thévenin's theorem, Norton's theorem, Superposition Theorem, etc.;

Chapter 4 - Circuit analysis is the procedure of determining the currents and voltages across electrical components of a network. This chapter has been carefully written to provide an easy understanding of the varied facets of circuit analysis techniques, such as network analysis, mesh analysis, super mesh transform, nodal analysis, distributed element model, large-signal model, symbolic circuit analysis, etc.;

Chapter 5 - The graphical representation of an electrical circuit is called a circuit diagram. Such diagrams are useful for the design, construction, operation and maintenance of electronic and electrical equipment. This chapter closely examines the fundamental aspects of the graphical representation of circuits and includes topics such as circuit diagram, difference between schematics and circuit diagram and And-inverter graph.

Finally, I would like to thank my fellow scholars who gave constructive feedback and my family members who supported me at every step.

Jasper Harrison

Circuit and its Types

A circuit is an electrical network with a closed loop that provides a return path for the current. The aim of this chapter is to explore the different aspects of circuits and their different types, such as electronic and electrical circuits, and series and parallel circuits.

Circuit

In the most fundamental sense, the term circuit refers to a physical or conceptual loop, often with an emphasis on resolving or closing that loop to accommodate various operations associated with electrical systems or other technologies.

A closed circuit has a complete path for current to flow. An open circuit doesn't, which means that it's not functional. If this is your first exposure to circuits, you might think that when a circuit is open, it's like an open door or gate that current can flow through. And when it's closed, it's like a shut door that current can't flow through.

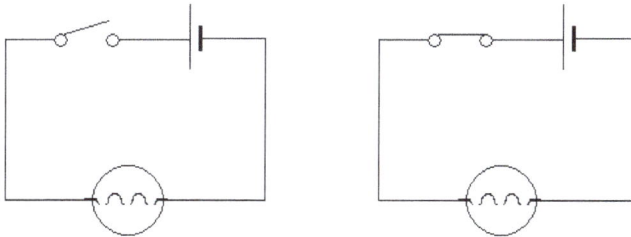

Series and Parallel Circuits

The Analog electronic circuit includes an analog signal with any continuously changeable signal. While working on an analog signal, an analog circuit alters the signal in some manner. Analog circuit can be used to convert the original signal into some other format such as a digital signal. Analog circuits may also modify signals in inadvertent ways like adding noise or distortion. Analog circuits are classified into two types, namely active analog circuits and passive analog circuits. An analog circuit uses an electrical power source to get the goals of a designer while Passive circuits use no external electrical power.

Series

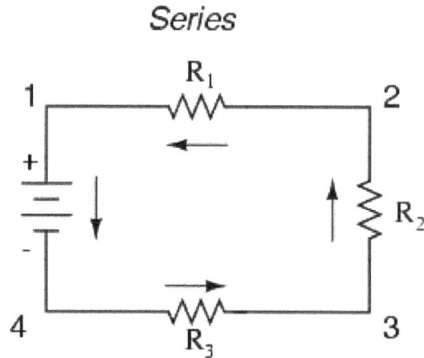

Here, we have three resistors (labeled R_1, R_2, and R_3), connected in a long chain from one terminal of the battery to the other. (It should be noted that the subscript labeling—those little numbers to the lower-right of the letter "R"—are unrelated to the resistor values in ohms. They serve only to identify one resistor from another.) The defining characteristic of a series circuit is that there is only one path for electrons to flow. In this circuit the electrons flow in a counter-clockwise direction, from point 4 to point 3 to point 2 to point 1 and back around to 4.

Now, let's look at the other type of circuit, a parallel configuration:

Parallel

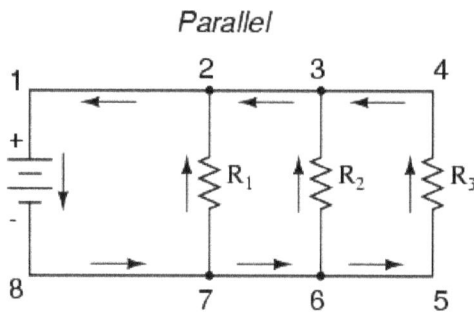

Again, we have three resistors, but this time they form more than one continuous path for electrons to flow. There's one path from 8 to 7 to 2 to 1 and back to 8 again. There's another from 8 to 7 to 6 to 3 to 2 to 1 and back to 8 again. And then there's a third path from 8 to 7 to 6 to 5 to 4 to 3 to 2 to 1 and back to 8 again. Each individual path (through R_1, R_2, and R_3) is called a *branch*.

The defining characteristic of a parallel circuit is that all components are connected between the same set of electrically common points. Looking at the schematic diagram, we see that points 1, 2, 3, and 4 are all electrically common. So are points 8, 7, 6, and 5. Note that all resistors as well as the battery are connected between these two sets of points.

And, of course, the complexity doesn't stop at simple series and parallel either! We can have circuits that are a combination of series and parallel, too:

Series-parallel

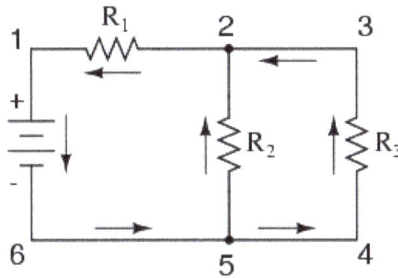

In this circuit, we have two loops for electrons to flow through: one from 6 to 5 to 2 to 1 and back to 6 again, and another from 6 to 5 to 4 to 3 to 2 to 1 and back to 6 again. Notice how both current paths go through R_1 (from point 2 to point 1). In this configuration, we'd say that R_2 and R_3 are in parallel with each other, while R_1 is in series with the parallel combination of R_2 and R_3.

This is just a preview of things to come. Don't worry! We'll explore all these circuit configurations in detail, one at a time.

The basic idea of a "series" connection is that components are connected end-to-end in a line to form a single path for electrons to flow:

Series connection

only one path for electrons to flow!

The basic idea of a "parallel" connection, on the other hand, is that all components are connected across each other's leads. In a purely parallel circuit, there are never more than two sets of electrically common points, no matter how many components are connected. There are many paths for electrons to flow, but only one voltage across all components:

Parallel connection

These points are electrically common

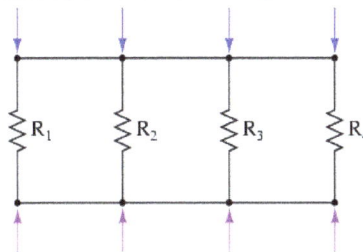

These points are electrically common

Series and parallel resistor configurations have very different electrical properties. We'll explore the properties of each configuration in the sections to come.

Electrical Circuit

An electric circuit is in many ways similar to your circulatory system. Your blood vessels, arteries, veins and capillaries are like the wires in a circuit. The blood vessels carry the flow of blood through your body. The wires in a circuit carry the electric current to various parts of an electrical or electronic system.

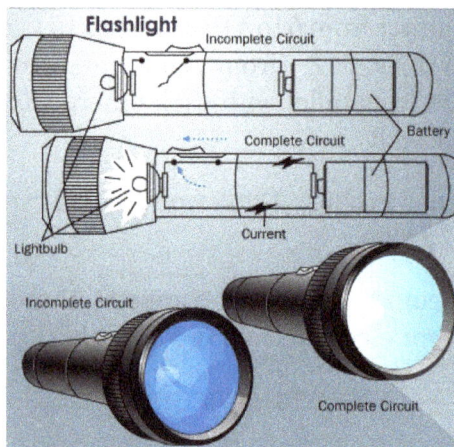

The circuit illustration above shows how the circuit of a flashlight works.

Your heart is the pump that drives the blood circulation in the body. It provides the force or pressure for blood to circulate. The blood circulating through the body supplies various organs, like your muscles, brain and digestive system. A battery or generator produces voltage the force that drives current through the circuit.

Take the simple case of an electric light. Two wires connect to the light. For electrons to do their job in producing light there must be a complete circuit so they can flow through the light bulb and then back out.

The diagram above shows a simple circuit of a flashlight with a battery at one end and a flashlight bulb at the other end. When the switch is off, a complete circuit will not exist, and there will be no current. When the switch is on, there will be a complete circuit and a flow of current resulting in the flashbulb emitting light.

Circuits can be huge power systems transmitting megawatts of power over a thousand miles -- or tiny microelectronic chips containing millions of transistors. This extraordinary shrinkage of electronic circuits made desktop computers possible. The new frontier promises to be Nano electronic circuits with device sizes in the nanometers (one-billionth of a meter).

The two basic types of electric are circuits:

- Power circuits transfer and control large amounts of electricity. Examples are power lines and residential and business wiring systems. The major components of power circuits are generators at one end and lighting systems, heating systems or household appliances at the other end. In between are power lines, transformers and circuit breakers.

- Electronic circuits process and transmit information. Think computers, radios, TVs, radars and cell phones.

Components of an Electrical Circuit

Cell

Cell is a device used to power electrical circuits. It has two terminals; positive and negative. The terminal marked negative is the source of electrons, that when connected to a circuit delivers energy. We can take the example of a normal torch battery. A battery is a combination of multiple cells. If we take a closer look at it, we can see positive and negative marks. The same marks are present on the torch we put our batteries into. When we insert the battery into our devices matching these signs, we complete a circuit and that is how the circuit gets its power. Once the reserve of excess electron inside the cells is over, the delivery of energy stops and the circuit breaks. Thus, the device connected to the circuit, such as a bulb, stops working. Batteries come in various shapes and sizes, starting from the miniature batteries used in hearing aids and watches, torch and mobile batteries to lead acid batteries used in cars and to power inverters.

Switch

Switch is a device that can break an electrical circuit by diverting the current from one conductor to another conductor or an insulator. These set of contacts are termed as open and closed. Open circuit means that the contacts are separated and the circuit is broken, so no current is flowing. Whereas, closed circuit means that the contacts are touching, the circuit is complete and the current is flowing. Switches are of many types and are used depending upon the device we are using. Generally we use them in our house for controlling the fans, bulbs, call bells, power switches for devices like refrigerators, washing machines etc.

Light Bulb

Light bulb is a device that produces light from electricity. Light bulbs turn the electricity to light by sending current through a thin wire called filament. The filament is usually made of tungsten, a material that emits light when electricity is passed through it. The emission of light is due to the high resistance offered by the material tungsten, which we will learn in higher classes. Apart from lighting, the light bulbs are used in electronic items as an indicator, traffic signals, indicator lights in cars etc.

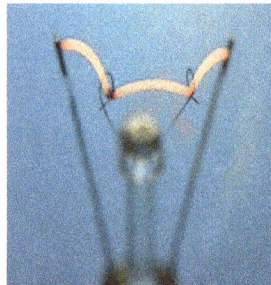

Connecting Wires

Wire is a flexible strand of metal, usually cylindrical in shape. Wires are used for establishing electrical conductivity between two devices of any electrical circuit. They possess negligible resistance to the passage of current. The wires are covered by insulated coating of different colors. The color codes are used to distinguish between, live, neutral and ground wire, which differs from one country to another.

Balanced Circuit

A balanced circuit is an electrical or electronic circuitry where the signal or power is transferred symmetrical in two wires. Unlike in the unbalanced circuit where the voltage is seen in one wire and the common, the output of the balanced circuit is usually two wires with one wire output at 180 degrees phase difference from the other wire. The balanced circuit is used for high-power or high-electromagnetic interference immunity.

For better noise performance, the simplest balanced circuit is a transformer output that has a secondary winding, which is either floating or has a center tap for common connection. Most audio and telecommunication baseband applications make use of balanced circuits to convey signals between equipment cabinets. Common balanced lines are twisted pairs that may be bundled as in the 25-pair cable that is also shielded to further take advantage of the electromagnetic interference immunity.

More sophisticated balanced circuits may use a bridge-type transistor amplifier stage or dual operational amplifier output. The resulting output at idle is 0 volts alternating current (VAC), which is the same as in the transformer output. An advantage of electronic drive over transformer coupled is the decreased size and weight of the resulting circuit. The frequency response of pure electronic and transformer-less circuits is usually superior to the transformer output counterparts.

Balanced circuits use miscellaneous components for better handling of electrical noise. In communication circuits, repeating coils are used for long spans of balanced lines to reduce noise such as the hum produced by electrical circuits and the transient cracking noise with lightning discharge. These devices also reduce noise from earth loop currents that are caused when alternating current (AC), which is supposed to run through the neutral wires, is mistakenly routed into ground cabling. This happens when the AC power termination has a neutral-to-ground connection in the equipment side instead of the AC mains power distribution cabinet.

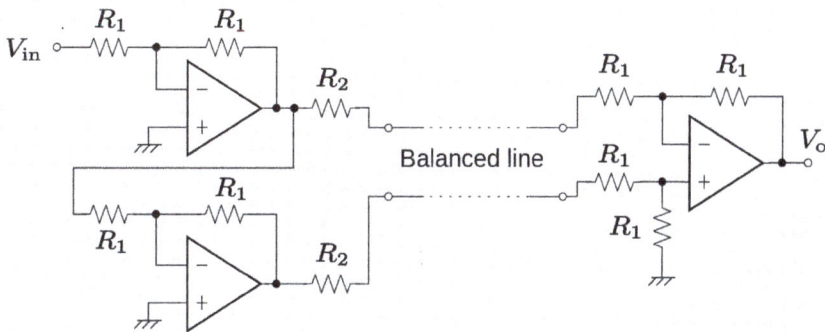

Unbalanced Circuit

An *unbalanced circuit* (or "unbalanced line") is a signal-carrying circuit with one electrical conductor and an overall metallic shield. Unbalanced connections are typically found on professional-level instrument inputs and outputs, and such consumer-level products as turntable RCA connections, headphone outputs, and the 3.5 mm (⅛ inch) connections used for microphone outputs, PC microphone inputs, Mac line-level inputs, and some camera inputs.

Electronic Circuit

An electronic circuit is a complete course of conductors through which current can travel. Circuits provide a path for current to flow. To be a circuit, this path must start and end at the same point. In other words, a circuit must form a loop. An electronic circuit and an electrical circuit has the same definition, but electronic circuits tend to be low voltage circuits.

For example, a simple circuit may include two components: a battery and a lamp. The circuit allows current to flow from the battery to the lamp, through the lamp, then back to the battery. Thus, the circuit forms a complete loop.

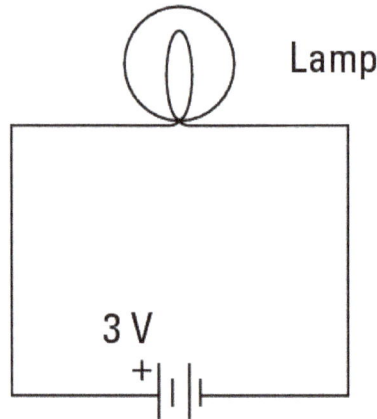

Lamp

3 V
+

Of course, circuits can be more complex. However, all circuits can be distilled down to three basic elements:

Voltage source: A voltage source causes current to flow like a battery, for instance.

Load: The load consumes power; it represents the actual work done by the circuit. Without the load, there's not much point in having a circuit.

The load can be as simple as a single light bulb. In complex circuits, the load is a combination of components, such as resistors, capacitors, transistors, and so on.

Conductive path: The conductive path provides a route through which current flows. This route begins at the voltage source, travels through the load, and then returns to the voltage source. This path must form a loop from the negative side of the voltage source to the positive side of the voltage source.

When a circuit is complete and forms a loop that allows current to flow, the circuit is called a closed circuit. If any part of the circuit is disconnected or disrupted so that a loop is not formed, current cannot flow. In that case, the circuit is called an open circuit.

Open circuit is an oxymoron. After all, the components must form a complete path to be considered a circuit. If the path is open, it isn't a circuit. Therefore, open circuit is most often used to describe a circuit that has become broken, either on purpose (by the use of a switch) or by some error, such as a loose connection or a damaged component.

Short circuit refers to a circuit that does not have a load. For example, if the lamp is connected to the circuit but a direct connection is present between the battery's negative terminal and its positive terminal, too.

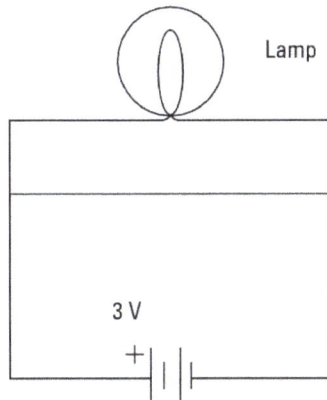

Current in a short circuit can flow at dangerously high levels. Short circuits can damage electronic components, cause a battery to explode, or maybe start a fire.

The short circuit illustrates an important point about electrical circuits: it is possible — common, even — for a circuit to have multiple pathways for current to flow. The current can flow through the lamp as well as through the path that connects the two battery terminals directly.

Current flows everywhere it can. If your circuit has two pathways through which current can flow, the current doesn't choose one over the other; it chooses both. However, not all paths are equal, so current doesn't flow equally through all paths.

For example, current will flow much more easily through the short circuit than it will through the lamp. Thus, the lamp will not glow because nearly all of the current will bypass the lamp in favor of the easier route through the short circuit. Even so, a small amount of current will flow through the lamp.

Analog Circuit

The Analog electronic circuit includes an analog signal with any continuously changeable signal. While working on an analog signal, an analog circuit alters the signal in some manner. Analog circuit can be used to convert the original signal into some other format such as a digital signal. Analog circuits may also modify signals in inadvertent ways like adding noise or distortion. Analog circuits are classified into two types, namely active analog circuits and passive analog circuits. An analog circuit uses an electrical power source to get the goals of a designer while Passive circuits use no external electrical power.

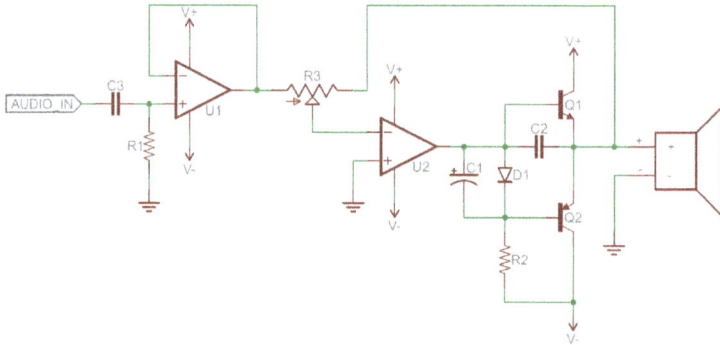

Analog Circuit

The working of an analog circuits can be done with normal waveforms, changing them. For ex: a microphone is used in an analog circuit which converts the sound waves into analog electric waves. For example, in an analog circuit a microphone converts sound waves into similar or analog electric waves. These signals can be stored on the strip, improved in an analog amplifier and transformed back to related sound waves by a speaker.

Digital Circuit

A digital circuit is a circuit where the signal should be one of two discrete levels. Each level is interpreted as one of two different states (for instance, 0 or 1). These circuits built with transistors to make logic gates in order to execute Boolean logic operation. This logic is the base of digital electronics & computer processing. Digital circuits are less vulnerable to degradation in excellence than analog circuits. It is also simpler to execute error detection and rectification with digital signals. To make the routine process of designing these circuits, designers use EDA (electronic design automation) tools, a kind of software that develops the logic in a digital circuit.

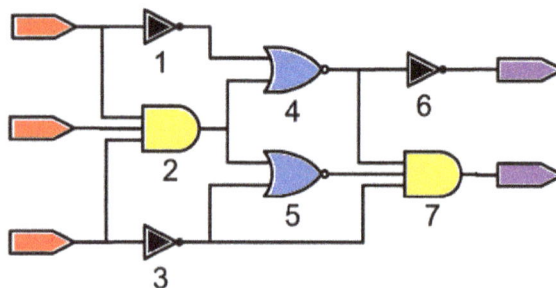

Digital Circuit

A digital circuit is used to alter the waves into pulse waves. It measures a waveform thousand of times every second and stores the data in binary form. For instance, after 12msecs, a signal might be 2.4 volts high & after 14msecs, it might be at 2.6 volts. This circuit changes the volts and times into binary data, and the waves become a series of 1's and 0's. When the circuit has to generate sound from a speaker, it produces an o/p signal which is at 2.4 V after 12msec and at 2.6 V after 14msec, similar to the original wave.

Difference between Analog and Digital Circuit

The main differences between analog circuit and digital circuits mainly includes the following:

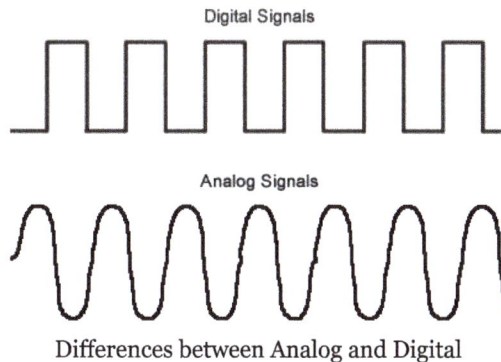

Digital Signals

Analog Signals

Differences between Analog and Digital

- Analog circuits operate on analog signals commonly known as continuous valued signals.

- Digital circuits function on signals that exist simply at 2 levels, i.e. Zeros and ones.

- The designing of an analog circuit is difficult since every component has to be positioned by hand for designing the circuits.

- Digital circuits are very simple to design since the technique of automation can be useful at a variety of levels of circuit design.

- No change of i/p signals is necessary before processing, the circuit straightly executes different logical operations and generates an analog o/p.

- In digital circuits, the i/p signals change from analog to digital (A/D) form before it is processed, that is the digital circuit is accomplished by processing digital signals only, and generates o/p which is again changed back from digital to analog signals (D/A) so that the o/p gives relevant results that can be understood by individuals.

- Analog circuits are typically routine made and they don't have flexibility.

- Digital circuits have a high degree of elasticity.

Asynchronous Circuit

Asynchronous sequential logic is not synchronized by a clock signal; the outputs of the circuit change directly in response to changes in inputs. The advantage of asynchronous logic is that it can be faster than synchronous logic, because the circuit doesn't have to wait for a clock signal to process inputs. The speed of the device is potentially limited only by the propagation delays of the logic gates used.

However, asynchronous logic is more difficult to design and is subject to problems not encountered in synchronous designs. The main problem is that digital memory elements are sensitive to the order that their input signals arrive; if two signals arrive at a logic gate at almost the same time, which state the circuit goes into can depend on which signal gets to the gate first. Therefore the circuit can go into the wrong state, depending on small differences in the propagation delays of the logic gates. This is called a race condition. This problem is not as severe in synchronous circuits because the outputs of the memory elements only change at each clock pulse. The interval between clock signals is designed to be long enough to allow the outputs of the memory elements to "settle" so they are not changing when the next clock comes. Therefore the only timing problems are due to "asynchronous inputs"; inputs to the circuit from other systems which are not synchronized to the clock signal.

Asynchronous sequential circuits are typically used only in a few critical parts of otherwise synchronous systems where speed is at a premium, such as parts of microprocessors and digital signal processing circuits.

Synchronous Circuit

In a synchronous circuit, an electronic oscillator called a clock generates a sequence of repetitive pulses called the clock signal which is distributed to all the memory elements in the circuit. The basic memory element in sequential logic is the flip-flop. The output of each flip-flop only changes when triggered by the clock pulse, so changes to the logic signals throughout the circuit all begin at the same time, at regular intervals, synchronized by the clock.

The output of all the storage elements (flip-flops) in the circuit at any given time, the binary data they contain, is called the state of the circuit. The state of a synchronous circuit only changes on clock pulses. At each cycle, the next state is determined by the current state and the value of the input signals when the clock pulse occurs.

The main advantage of synchronous logic is its simplicity. The logic gates which perform the operations on the data require a finite amount of time to respond to changes to their inputs. This is called propagation delay. The interval between clock pulses must be long enough so that all the logic gates have time to respond to the changes and their outputs "settle" to stable logic values, before the next clock pulse occurs. As long as this condition is met (ignoring certain other details) the circuit is guaranteed to be stable and reliable. This determines the maximum operating speed of a synchronous circuit.

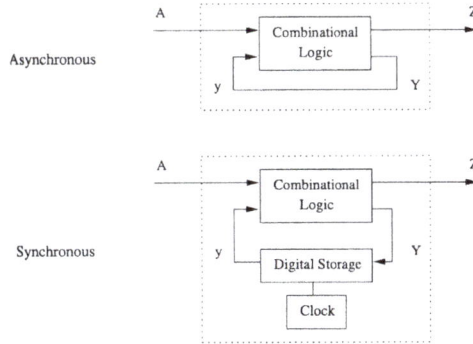

Integrated Circuit

An Integrated circuit is a special component which is fabricated with thousands of transistors, resistors, diodes and other electronic components on a tiny silicon chip. These are the building blocks of current electronic devices like cell phones, computers, etc. These can be analog or digital integrated circuits. Mostly used ICs in electronic circuits are Op-amps, timers, comparators, switches ICs and so on. These can be classified as linear and nonlinear ICs depending on its application.

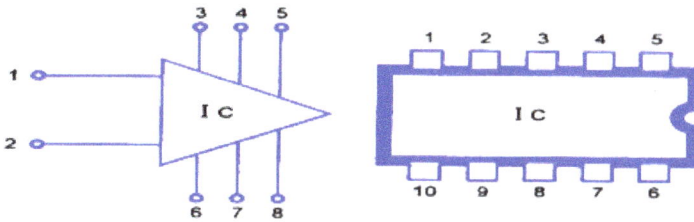

Integrated Circuits

There are different types of ICs; classification of Integrated Circuits is done based on various criteria. A few types of ICs in a system are shown in the below figure with their names in a tree format.

Different Types of ICs

Based on the intended application, the IC are classified as analog integrated circuits, digital integrated circuits and mixed integrated circuits.

Digital Integrated Circuits

The integrated circuits that operate only at a few defined levels instead of operating over all levels of signal amplitude are called as Digital ICs and these are designed by using multiple numbers of digital logic gates, multiplexers, flip flops and other electronic components of circuits. These logic gates work with binary input data or digital input data, such as 0 (low or false or logic 0) and 1 (high or true or logic 1).

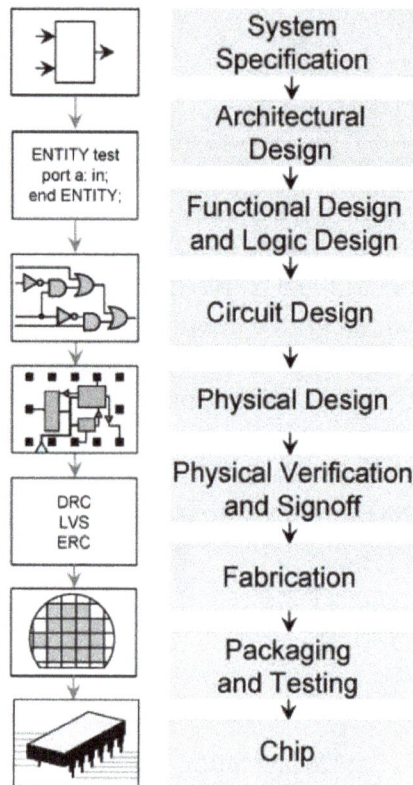

ENTITY test
port a: in;
end ENTITY;

DRC
LVS
ERC

System Specification
↓
Architectural Design
↓
Functional Design and Logic Design
↓
Circuit Design
↓
Physical Design
↓
Physical Verification and Signoff
↓
Fabrication
↓
Packaging and Testing
↓
Chip

Digital Integrated Circuits

The above figure shows the steps involved in designing a typical digital integrated circuits. These digital ICs are frequently used in the computers, microprocessors, digital signal processors, computer networks and frequency counters. There are different types of digital ICs or types of digital integrated circuits, such as programmable ICs, memory chips, logic ICs, power management ICs and interface ICs.

Analog Integrated Circuits

The integrated circuits that operate over a continuous range of signal are called as Analog ICs. These are subdivided as linear Integrated Circuits (Linear ICs) and Radio Frequency

Integrated Circuits (RF ICs). In fact, the relationship between the voltage and current maybe nonlinear in some cases over a long range of the continuous analog signal.

Analog Integrated Circuits

The frequently used analog IC is an operational amplifier or simply called as an op-amp, similar to the differential amplifier, but possesses a very high voltage gain. It consists of very less number of transistors compared to the digital ICs, and, for developing analog application specific integrated circuits (analog ASICs), computerized simulation tools are used.

Mixed Integrated Circuits

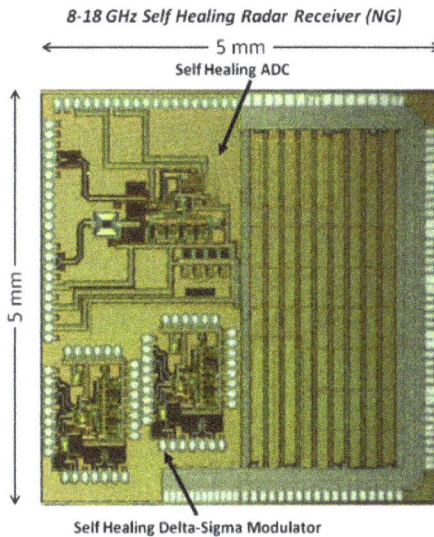

Mixed Integrated Circuits

The integrated circuits that are obtained by the combination of analog and digital ICs on a single chip are called as Mixed ICs. These ICs functions as Digital to Analog converters, Analog to Digital converters (D/A and A/D converters) and clock/timing ICs.

The circuit depicted in the above figure is an example of mixed integrated circuit which is a photograph of the 8 to 18 GHz self healing radar receiver.

This mixed-signal Systems-on-a-chip is a result of advances in the integration technology, which enabled to integrate digital, multiple analog and RF functions on a single chip.

General types of integrated circuits (ICs) include the following:

Logic Circuits

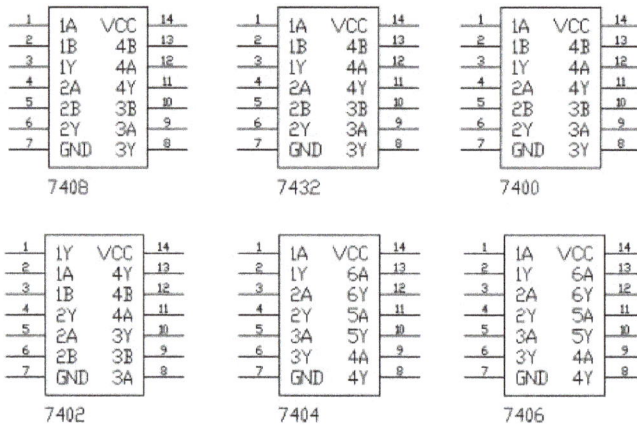

Logic Circuits

These ICs are designed using logic gates-that work with binary input and output (0 or 1). These are mostly used as decision makers. Based on the logic or truth table of the logic gates, all the logic gates connected in the IC give an output based on the circuit connected inside the IC- such that this output is used for performing a specific intended task. A few logic ICs are shown above.

Comparators

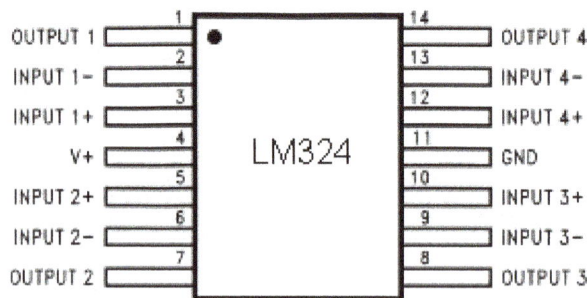

Comparators

The comparator ICs are used as comparators for comparing the inputs and then to produce an output based on the ICs' comparison.

Switching ICs

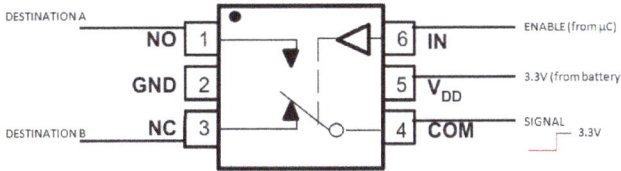

Switching ICs

Switches or Switching ICs are designed by using the transistors and are used for performing the switching operations. The above figure is an example showing an SPDT IC switch.

Audio Amplifiers

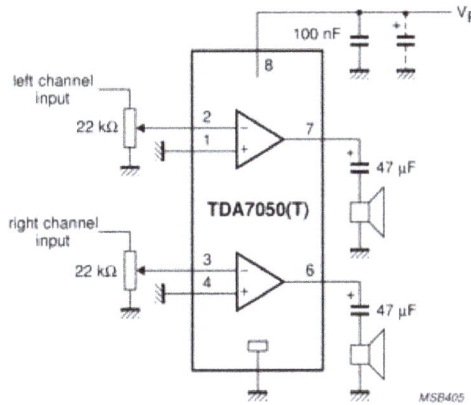

Audio amplifiers

The audio amplifiers are one of the many types of ICs, which are used for the amplification of the audio. These are generally used in the audio speakers, television circuits, and so on. The above circuit shows the low- voltage audio amplifier IC.

Operational Amplifiers

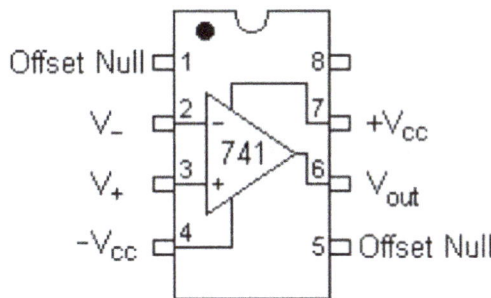

Operational amplifiers

The operational amplifiers are frequently used ICs, similar to the audio amplifiers which are used for the audio amplification. These op-amps are used for the amplification

purpose, and these ICs work similar to the transistor amplifier circuits. The pin configuration of the 741 op-amp IC is shown in the above figure.

Timer ICs

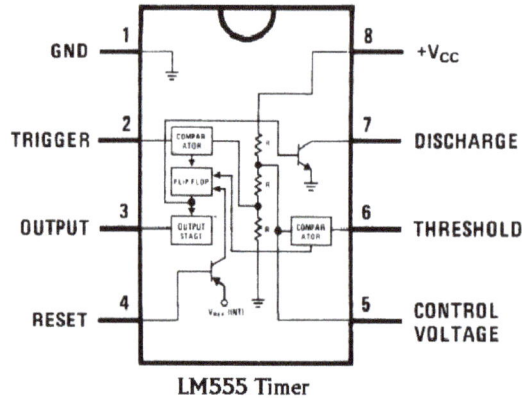

LM555 Timer

Timer ICs

Timers are special purpose integrated circuits used for the purpose of counting and to
keep a track of time in intended applications. The block diagram of the internal circuit
of the LM555 timer IC is shown in the above circuit.

Based on the number of components used (typically based on the number of transistors
used), they are as follows:

- Small-scale integration consists of only a few transistors (tens of transistors on
  a chip), these ICs played a critical role in early aerospace projects.

- Medium-scale integration consists of some hundreds of transistors on the IC chip
  developed in the 1960s and achieved better economy and advantages compared
  to the SSI ICs.

- Large-scale integration consists of thousands of transistors on the chip with
  almost the same economy as medium scale integration ICs. The first microprocessor, calculator chips and RAMs of 1Kbit developed in the 1970s had below
  four thousand transistors.

- Very large-scale integration consists of transistors from hundreds to several bil-
  lions in number.

- Ultra large-scale integration consists of transistors in excess of more than one
  million, and later wafer-scale integration (WSI), system on a chip (SoC) and
  three dimensional integrated circuit (3D-IC) were developed.

All these can be treated as generations of integrated technology. ICs are also classified
based on the fabrication process and packing technology. There are numerous types of

ICs among which, an IC will function as timer, counter, register, amplifier, oscillator, logic gate, adder, microprocessor, and so on.

The conventional Integrated circuits are reduced in practical usage, because of the invention of the Nano-electronics and the miniaturization of ICs being continued by this Nano-electronics technology. However, the conventional ICs are not yet replaced by nano-electronics but the usage of the conventional ICs is getting diminished partially.

LC Circuit

An LC circuit is also called a tank circuit, tuned circuit or resonant circuit, is an electric circuit built with a capacitor denoted by the letter 'C' and an inductor denoted by the letter 'L' connected together. These circuits are used for producing signals at a particular frequency or accepting a signal from a more composite signal at a particular frequency. LC circuits are basic electronics components in various electronic devices, especially in radio equipment used in circuits like tuners, filters, frequency mixers and oscillators. The main function of an LC circuit is generally to oscillate with minimum damping.

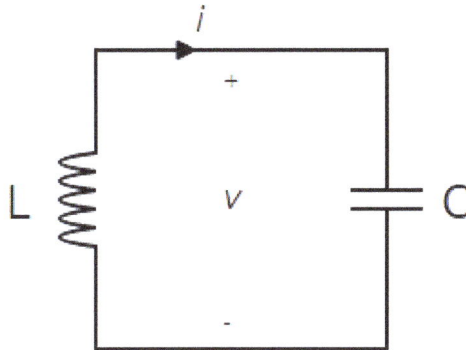

LC Circuit

Series LC Circuit Resonance

In the series LC circuit configuration, the capacitor 'C' and inductor 'L' both are connected in series that is shown in the following circuit. The sum of the voltage across the capacitor and inductor is simply the sum of the whole voltage across the open terminals. The flow of current in the +Ve terminal of the LC circuit is equal to the current through both the inductor (L) and the capacitor (C),

$$v = v_L + v_C$$

$$i = i_L = i_C$$

When the 'XL' inductive reactance magnitude increases, then the frequency also increases. In the same way while 'XC' capacitive reactance magnitude decreases, then the frequency decreases.

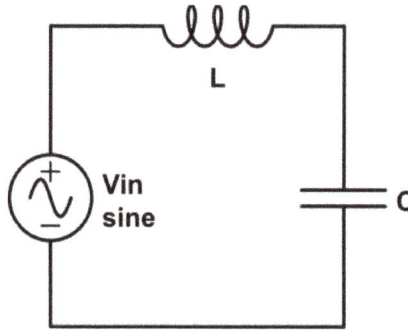

Series LC Circuit Resonance

At one specific frequency, the two reactances XL and XC are the same in magnitude but reverse in sign. So this frequency is called the resonant frequency which is denoted by for the LC circuit.

Therefore, at resonance,

$$X_L = -X_C$$

$$\omega_L = 1/\omega_C$$

$$\omega = \omega_0 = 1/\sqrt{LC}$$

Which is termed as the resonant angular frequency of the circuit. Changing angular frequency into frequency, the following formula is used,

$$fo = \omega o/2\pi \sqrt{LC}$$

In a series resonance LC circuit configuration, the two resonances XC and XL cancel each other out. In actual, rather than ideal components, the flow of current is opposed, generally by the resistance of the windings of the coil. Therefore, the current supplied to the circuit is a max at resonance.

An acceptance circuit is defined as when the In the Lt f<fo is the maximum and the impedance of the circuit is minimized.

For f<fo, $X_L << (-X_C)$. Thus, the circuit is capacitive

For f<fo, $X_L >> (-X_C)$. Thus, the circuit is inductive

Parallel LC Circuit Resonance

In the parallel LC circuit configuration, the capacitor 'C' and inductor 'L' both are connected in parallel that is shown in the following circuit. The sum of the voltage across the capacitor and inductor is simply the sum of the whole voltage across the open terminals. The flow of current in the +Ve terminal of the LC circuit is equal to the current through both the inductor (L) and the capacitor (C),

$$v = v_L = v_C$$

$$i = i_L + i_C$$

Let the internal resistance 'R' of the coil. When two resonances XC and XL, the reactive branch branch currents are the same and opposed. Therefore, they cancel out each other to give the smallest amount of current in the key line. When the total current is minimum in this state, then the total impedance is max. The resonant frequency is given by,

$$fo = \omega o / 2\pi = 1/2\pi \sqrt{LC}$$

Note that the current of any reactive branch is not minimum at resonance, but each is given individually by separating source voltage 'V' by reactance 'Z'.

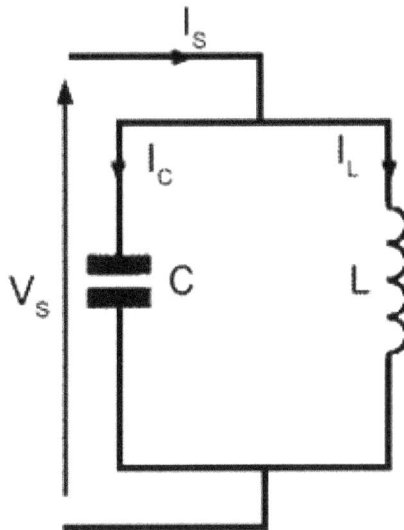

Parallel LC Circuit Resonance

Hence, according to Ohm's law $I=V/Z$

A rejector circuit can be defined as, when the line current is minimum and total impedance is max at fo, circuit is inductive when below fo and the circuit is capacitive when above fo.

Applications of LC Circuit:

- The applications of the resonance of the series and parallel LC circuits mainly involve in communications systems and signal processing.

- The common application of an LC circuit is, tuning radio TXs and RXs. For instance, when we tune a radio to an exact station, then the circuit will set at resonance for that specific carrier frequency.

- A series resonant LC circuit is used to provide voltage magnification.

- A parallel resonant LC circuit is used to provide current magnification and also used in the RF amplifier circuits as the load impedance, the amplifier's gain is max at the resonant frequency.

- Both series and parallel resonant LC circuits are used in induction heating.

- These circuits perform as electronic resonators, which are an essential component in various applications like amplifiers, oscillators, filters, tuners, mixers, graphic tablets, contactless cards and security tags X_L and X_C.

Thus, this is all about LC circuit, operation of series and parallel resonance circuits and its applications.

RC Circuit

The combination of a pure resistance R in ohms and pure capacitance C in Farads is called RC circuit. The capacitor store energy and the resistor connect in series with the capacitor control the charging and discharging of a capacitor. The RC circuit is used in camera flashes, pacemaker, timing circuit etc.

The RC signal filters the signals by blocking some frequencies and allowing others to pass through it. It is also called first order RC circuit and is used to filter the signals by passing some frequencies and blocking others. The RC filters are mostly use for selecting signals and for rejecting noise.

The high-pass filter and the low-pass filter are the most common type of RC filters. The high pass filter passes the frequency greater than the fixed cutoff frequency and blocks the frequency lower than the fixed cutoff frequency. Similarly, the low-pass filter allows the frequency lower than the fixed cutoff frequency and attenuates the frequency higher than the fixed cutoff frequency.

RC Series Circuit

A circuit containing resistance and capacitance connected in series together is called an RC series circuit.

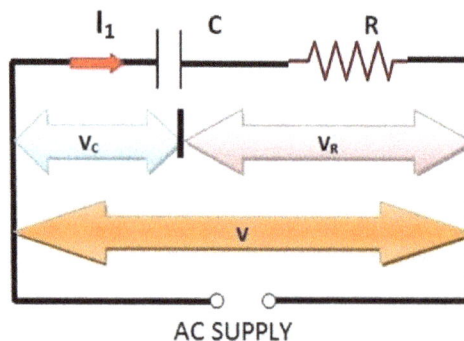

AC SUPPLY

Steps to draw a phasor diagram for an RC circuit

1. Current I is taken as reference

2. Voltage drop in resistance is (V_R).

   V_R = IR is drawn in phase with the current I

3. Voltage drop in capacitive reactance is (V_C).

   V_C = IXC and is drawn 90° behind the current (as current leads the voltage by 90° in pure capacitive load circuit).

4. The vector sum of the two voltage drops is equal to the applied voltage (V).

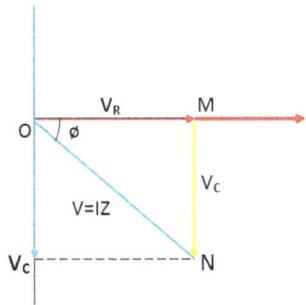

$V_R = IR$

$V_C = IX_C \ where \ X_C = \dfrac{1}{2\pi fc}$

Now in right angle triangle OMN

$$V = \sqrt{\left(V_R\right)^2 + \left(V_C\right)^2}$$
$$V = \sqrt{\left(IR\right)^2 + \left(IX_c\right)^2}$$
$$V = I/\sqrt{R^2 + X^2 c}$$
$$I = V/\sqrt{R^2 + X^2 c} = V/Z$$
$$Where \ Z = \sqrt{R^2 + X^2 c}$$

Z is known as the impedance of the circuit and is defined as the total opposition offered to the flow of current in an RC series circuit. It is measured in ohms.

RL Circuit

RL Circuit refers to a circuit having combination of resistance(s) and inductor(s). They are commonly used in chokes of luminescent tubes. In an A.C. circuit, inductors helps in reducing voltage, without the loss of energy. Due to the inductive reactance, the higher the AC frequency, the greater the impeadence of the inductor. Under DC conditions, an inductor acts as a static resistance.

Like RC circuit, with one resistor and one coil can be connected to form a low pass filter or a high pass filter. A high-pass filter allows frequencies above the cut-off frequency to pass, while a low-pass filter allows frequencies beneath the cut-off frequency to pass. The arrangement of the resistor and the capacitor is what determines their behaviour.

Note that at a particular frequencly, called the cut-off frequency, the Inductive Reactance is equal to the Resistance value. (There is also an associated phase shift of 45 degrees).

$$R = X_L$$

Substituting $X_L = 2\pi fL$

we then have,

$$R = 2\pi fL$$

The cut-off frequency, defined as the frequency at which the signal power is attenuated by 50% (or 3.01 dB), is a function of the resistive and capacitive values. We can rearrange the above formula to solve for f as follows:

$$f_{cut-off} = \frac{R}{2\pi L}$$

RL Series

A circuit of 2 components a resistor and an inductor connected in series,

$$V_L + V_R = 0$$

$$L\frac{d}{dt}i(t) + Ri(t) = 0$$

$$\frac{d}{dt}i(t) = -\frac{R}{L}i(t)$$

$$\int \frac{di(t)}{i(t)} = -\int \frac{R}{L}dt$$

$$Lni(t) = -\frac{R}{L}t + c$$

$$i(t) = e^{-\frac{R}{L}t+c}$$

$$i(t) = Ae^{-\frac{R}{L}t}$$

RL Filters

High Pass Filter

When the inductor is in parallel with the load while the resistor is in series with the inductor and load, this creates a high pass filter.

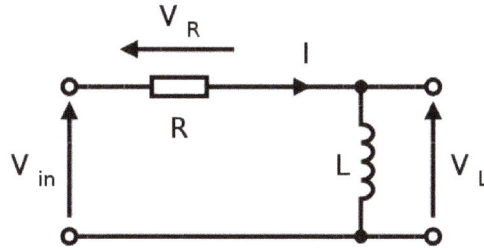

High pass filter has a transfer function,

$$H(j\omega) = \frac{v_o}{v_i} = \frac{j\omega T}{1 + j\omega T}$$

$$T = \frac{L}{R}$$

Frequency response of High pass filter,

$$\omega = 0. v_o = 0$$

$$\omega = \omega_o. v_o = \frac{v_i}{2}$$

$$\omega = \infty. v_o = v_i$$

Cut off frequency, ω_o, frequency at which $v_o = \frac{1}{2}v_i$

$$\omega_o = \frac{1}{T}$$

Low Pass Filter

When the resistor is in parallel with the load while the inductor is in series with the resistor and load, a low pass filter is created.

Low pass filter has a transfer function,

$$H(j\omega) = \frac{v_o}{v_i} = \frac{1}{1 + j\omega T}$$

$$T = \frac{L}{R} = RC$$

Frequency response of Low pass filter,

$$\omega = 0. v_o = 0$$

$$\omega = \omega_o . v_o = \frac{v_i}{2}$$

$$\omega = 00. v_o = v_i$$

Cut off frequency, ω_o, frequency at which $v_o = \frac{1}{2} v_i$

$$\omega_o = \frac{1}{T}$$

A single RL circuit creates a filter with a 20.0 dB/decade, or 6.02 dB/octave, slope.

Difference between RC and RL Circuit

Comparison Chart

Parameters	RC Circuit	RL Circuit
Definition	The RC circuit is a series connection of resistance and capacitance, this circuit stores energy in the form of electric field.	The RL circuit is a series combination of resistance and inductance which stores energy in the form of magnetic energy.
Power Dissipation	High Power Dissipation.	Low Power Dissipation.
Filtering of Signals	Appropriate for filtering of low power signals.	Appropriate for filtering of high power signals.
Size and Cost	Small in size, light weight and cheap.	Inductors are large in size and thus RL circuits are bulky and expensive.
Noise	RC circuit generated low noise or negligible noise.	RL circuit consists of inductors which creates magnetic field which creates hysteresis and noise in the circuit.

RLC Circuit

In RLC circuit, the most fundamental elements like resistor, inductor and capacitor are connected across a voltage supply. All these elements are linear and passive in nature; i.e. they consume energy rather than producing it and these elements have a linear relationship between voltage and current. There are number of ways of connecting these elements across voltage supply, but the most common method is to connect these elements either in series or in parallel. The RLC circuit exhibits the property of resonance in same way as LC circuit exhibits, but in this circuit the oscillation dies out quickly as compared to LC circuit due to the presence of resistor in the circuit.

Series RLC Circuit

When a resistor, inductor and capacitor are connected in series with the voltage supply, the circuit so formed is called series RLC circuit.

Since all these components are connected in series, the current in each element remains the same,

$$I_R = I_L = I_C = I(t)$$

where $I(t) = I_M \sin \omega t$

Let V_R be the voltage across resistor, R.

V_L be the voltage across inductor, L.

V_C be the voltage across capacitor, C.

X_L be the inductive reactance.

X_C be the capacitive reactance.

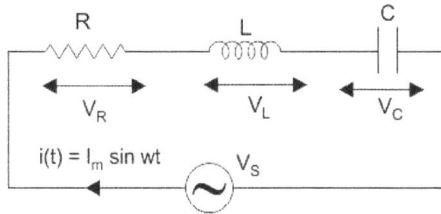

The total voltage in RLC circuit is not equal to algebraic sum of voltages across the resistor, the inductor and the capacitor; but it is a vector sum because, in case of resistor the voltage is in-phase with the current, for inductor the voltage leads the current by 90° and for capacitor, the voltage lags behind the current by 90° So, voltages in each component are not in phase with each other; so they cannot be added arithmetically. The figure below shows the phasor diagram of series RLC circuit. For drawing the phasor diagram for RLC series circuit, the current is taken as reference because, in series circuit the current in each element remains the same and the corresponding voltage vectors for each component are drawn in reference to common current vector,

$$V_s^2 = V_R^2 + (V_L - V_C)^2 \ (if V_L \rangle V_C)$$
$$V_s^2 = V_R^2 + (V_L - V_C)^2 \ (if V_L \langle V_C)$$
$$Where \ V_R = IR, V_L = IX_L, V_C = IX_C$$

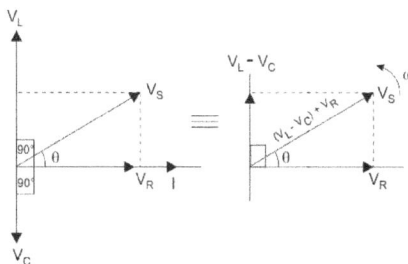

The Impedance for a Series RLC Circuit

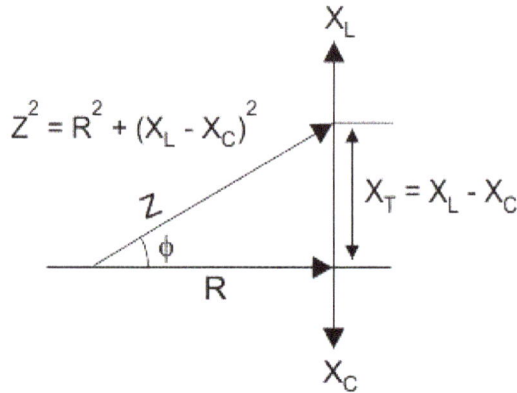

$$Z^2 = R^2 + (X_L - X_C)^2$$

$$X_T = X_L - X_C$$

The impedance Z of a series RLC circuit is defined as opposition to the flow of current due circuit resistance R, inductive reactance, X_L and capacitive reactance, X_C. If the inductive reactance is greater than the capacitive reactance i.e $X_L > X_C$, then the RLC circuit has lagging phase angle and if the capacitive reactance is greater than the inductive reactance i.e $X_C > X_L$ then, the RLC circuit have leading phase angle and if both inductive and capacitive are same i.e $X_L = X_C$ then circuit will behave as purely resistive circuit.

We know that

$$V_s^2 = V_R^2 + \left(V_L - V_C\right)^2$$

Where,

$$V_R = IR, V_L = IX_L, V_C = IX_C$$

Substituting the values

$$V_S^2 = IR^2 + \left(IX_L - IX_C\right)^2$$

$$V_S = \sqrt{R^2 + \left(X_L - X_C\right)^2} \text{ or impedance } Z = \sqrt{R^2 + \left(X_L - X_C\right)^2}$$

Parallel RLC Circuit

In parallel RLC Circuit the resistor, inductor and capacitor are connected in parallel across a voltage supply. The parallel RLC circuit is exactly opposite to the series RLC circuit. The applied voltage remains the same across all components and the supply current gets divided. The total current drawn from the supply is not equal to mathematical sum of the current flowing in the individual component, but it is equal to its vector sum of all the currents, as the current flowing in resistor, inductor and capacitor are not in the same phase with each other; so they cannot be added arithmetically.

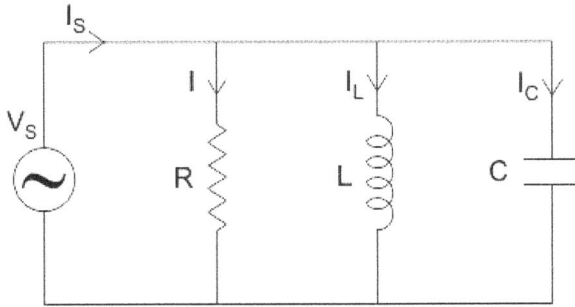

Phasor diagram of parallel RLC circuit, I_R is the current flowing in the resistor, R in amps.

I_C is the current flowing in the capacitor, C in amps.

I_L is the current flowing in the inductor, L in amps.

I_s is the supply current in amps.

In the parallel RLC circuit, all the components are connected in parallel; so the voltage across each element is same. Therefore, for drawing phasor diagram, take voltage as reference vector and all the other currents i.e I_R, I_C, I_L are drawn relative to this voltage vector. The current through each element can be found using Kirchhoff's Current Law, which states that the sum of currents entering a junction or node is equal to the sum of current leaving that node.

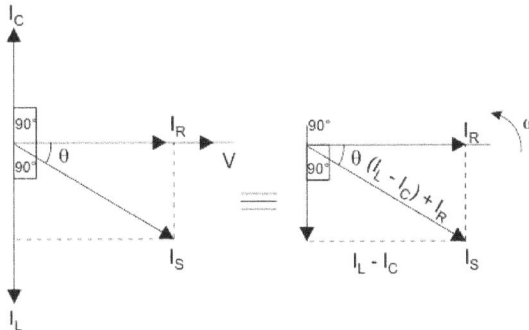

$$I_s^2 = I_R^2 + \left(I_L - I_C\right)^2$$

$$Now, I_R = \frac{V}{R}, I_C = \frac{V}{X_C} \text{ and } I_L = \frac{V}{X_L}$$

$$I_S = \sqrt{ + \left(\frac{V}{X_L} - \frac{V}{X_C}\right)^2}$$

$$So, admitace, \frac{1}{Z} = \frac{I_s}{V} = Y = \sqrt{\frac{1}{R^2} + \left(\frac{V}{X_L} - \frac{V}{X_C}\right)^2}$$

As shown above in the equation of impedance, Z of a parallel RLC circuit; each element has reciprocal of impedance (1/Z) i.e. admittance, Y. So in parallel RLC circuit, it is convenient to use admittance instead of impedance.

Resonance in RLC Circuit

In a circuit containing inductor and capacitor, the energy is stored in two different ways:

1. When a current flows in a inductor, energy is stored in magnetic field.

2. When a capacitor is charged, energy is stored in static electric field.

The magnetic field in the inductor is built by the current, which gets provided by the discharging capacitor. Similarly, the capacitor is charged by the current produced by collapsing magnetic field of inductor and this process continues on and on, causing electrical energy to oscillate between the magnetic field and the electric field. In some cases at certain frequency called resonant frequency, the inductive reactance of the circuit becomes equal to capacitive reactance which causes the electrical energy to oscillate between the electric field of the capacitor and magnetic field of the inductor. This forms a harmonic oscillator for current. In RLC circuit, the presence of resistor causes these oscillation s to die out over period of time and it is called as the damping effect of resistor.

Formula for Resonant Frequency

During resonance, at certain frequency called resonant frequency, f_r.

$$X_L = X_C$$

$$We\ know\ that,\ X_L = 2\pi fL\ and\ X_c = \frac{1}{2\pi fC}$$

$$Therefore\ at\ resonant\ frequency,\ f_r : 2\pi f_r L = \frac{1}{2\pi fC}$$

$$or\ f = \frac{1}{2\pi\sqrt{LC}}$$

When resonance occurs, the inductive reactance of the circuit becomes equal to capacitive reactance, which causes the circuit impedance to be minimum in case of series RLC circuit; but when resistor, inductor and capacitor are connected in parallel, the circuit impedance becomes maximum, so the parallel RLC circuit is sometimes called as anti resonator.

Difference between Series RLC Circuit and Parallel RLC Circuit

S.no	RLC Series Circuit	RLC Parallel Circuit
1	Resistor, inductor and capacitor are connected in series	Resistor, inductor and capacitor are connected in parallel
2	Current is same in each element	Current is different in all elements and the total current is equal to vector sum of each branch of current i.e $I_s^2 = I_R^2 + (I_C - I_L)^2$
3	Voltage across all the elements is different and the total voltage is equal to the vector sum of voltages across each component i.e $V_s^2 = V_R^2 + (V_L - V_C)^2$	Voltage across each element remains the same
4	For drawing phasor diagram, current is taken as reference vector	For drawing phasor diagram, voltage is taken as reference vector
5	Voltage across each element is given by : $V_R = IR,\ V_L = IX_L,\ V_C = IX_C$	Current in each element is given by: $I_R = V/R,\ I_C = V/X_C,\ I_L = V/X_L$
6	Its more convenient to use impedance for calculations	Its more convenient to use admittance for calculations
7	At resonance , when $X_L = X_C$, the circuit has minimum impedance	At resonance, when $X_L = X_C$, the circuit has maximum impedance

Equation of RLC Circuit

Consider a RLC circuit having resistor R, inductor L, and capacitor C connected in series and are driven by a voltage source V. Let Q be the charge on the capacitor and the current flowing in the circuit is I. Apply Kirchhoff's voltage law,

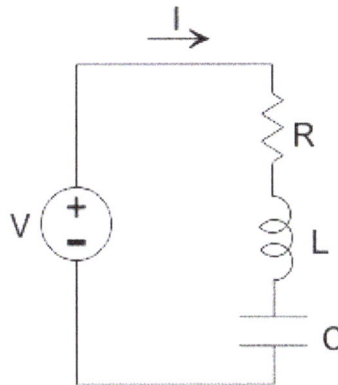

$$L.I'(t) + Q.I(t) + \frac{1}{C}Q(t) = V(t)$$

In this equation; resistance, inductance, capacitance and voltage are known quantities but current and charge are unknown quantities. We know that an current is a rate of electric charge flowing, so it is given by,

$$\frac{dQ}{dt}(t) = I(t) \text{ or } I(t) = Q(t)$$

Differentiating again I'(t) = Q" (t)

$$L.Q''(t) + R.Q'(t) + \frac{1}{C} = V(t)$$

Differentiating the above equation with respect to 't' we get,

$$L.I'(t) + Q.I(t) + \frac{1}{C}I(t) = V'(t)$$

Now at time t = 0, V(0) = 0 and at time t = t, V(t) = $E_o$sinωt

Differentiating with respect to 't' we get V'(t) = ωE_ocosωt

Substitute the value of V'(t) in above equation

$$L.I''(t) + R.I'(t) + \frac{1}{C}I(t) = \omega E_o \cos \omega t$$

Let us say that the solution of this equation is I_p(t) = Asin(ωt - ǿ) and if I_p(t) is a solution of above equation then it must satisfy this equation,

$$L.I_p(t) + R.I_p(t) + \frac{1}{C}I_p(t) = \omega E_o \cos \omega t$$

Now substitute the value of I_p(t) and differentiate it we get,

$$-L\omega 2A \sin(\omega t - \phi) + R\omega A \cos(\omega t - \phi) + \frac{1}{C}A \sin(\omega t - \phi) = \omega E_o \cos \omega t$$

$$-L\omega 2A \sin(\omega t - \phi) + R\omega A \cos(\omega t - \phi) + \frac{1}{C}A \sin(\omega t - \phi) = \omega E_o \cos(\omega t - \phi + \phi)$$

Apply the formula of cos (A + B) and combine similar terms we get,

$$\left(\frac{1}{C} - L\omega^2\right)A \sin(\omega t - \phi) + R\omega A \cos(\omega t - \phi)$$
$$= \omega E_o \cos \phi \cos(\omega t - \phi) - \omega E_o \sin \phi \cos(\omega t - \phi)$$

Match the coefficient of sin(ωt - φ) and cos(ωt - φ) on both sides we get,

$$\left(-\frac{1}{C}+2L\omega\right)A = \omega E_{\circ} \sin\phi \text{ and } R\omega A = \omega E_{\circ} \cos\phi$$

Now we have two equations and two unknowns i.e φ and A, and by dividing the above two equations we get,

$$\tan\phi = \frac{-\frac{1}{C}+2L\omega}{R\omega}$$

Squaring and adding above equation, we get,

$$A\sqrt{\left(-\frac{1}{C}+2L\omega\right)^2+\left(R\omega\right)^2} = \omega E_{\circ}$$

$$or \ A = \frac{\omega E_{\circ}}{\sqrt{\left(-\frac{1}{C}+2L\omega\right)^2+\left(R\omega\right)^2}}$$

Analysis of RLC Circuit Using Laplace Transformation

Step 1: Draw a phasor diagram for given circuit.

Step 2: Use Kirchhoff's voltage law in RLC series circuit and current law in RLC parallel circuit to form differential equations in the time-domain.

Step 3: Use Laplace transformation to convert these differential equations from time-domain into the s-domain.

Step 4: For finding unknown variables, solve these equations.

Step 5: Apply inverse Laplace transformation to convert back equations from s-domain into time domain.

Applications of RLC Circuit

It is used as low-pass filter, high-pass filter, band-pass filter, band-stop filter, voltage multiplier and oscillator circuit. It is used for tuning radio or audio receiver.

References

- What-are-series-and-parallel-circuits, direct-current: allaboutcircuits.com, Retrieved 29 July 2018
- Circuit-component, physics: byjus.com, Retrieved 19 May 2018
- What-is-a-balanced-circuit: wisegeek.com, Retrieved 25 March 2018

- What-is-an-electronic-circuit, components, electronics: dummies.com, Retrieved April 2018

- Synchronous-amp-Asynchronous-Sequential-Circuits-268121: edaboard.com, Retrieved 15 May 2018

- Major-electronic-components: elprocus.com, Retrieved 11 April 2018

- Series-and-parallel-lc-circuit-resonance: elprocus.com, Retrieved 31 March 2018

- What-is-an-rc-resistor-capacitor-circuit: circuitglobe.com, Retrieved 19 June 2018

Components of a Circuit

Any discrete electrical device that is used in an electronic system or circuit to affect the motion of electrons and their associated fields is termed as an electrical component. Some common circuit components are capacitors, resistors, inductors, voltage sources, switches, etc. which have been extensively covered in this chapter.

An electronic circuit comprises of various types of components, which are classified into two types: active components like transistors, diodes, IC's; and passive components like capacitors, resistors, inductors, etc.

Resistors

A resistor is a component or device designed to have a know value of resistance. OR, those components and devices which are specially designed to have a certain amount of resistance and used to oppose or limit the electric current are called resistors.

Resistance of a resistor depends on their length (l), resistivity (ρ) and its cross sectional area (a) which is also known as laws of resistance R = ρ (l/a).

Symbols of Different Types of Resistors. IEEE & IEC symbols of Resistors,

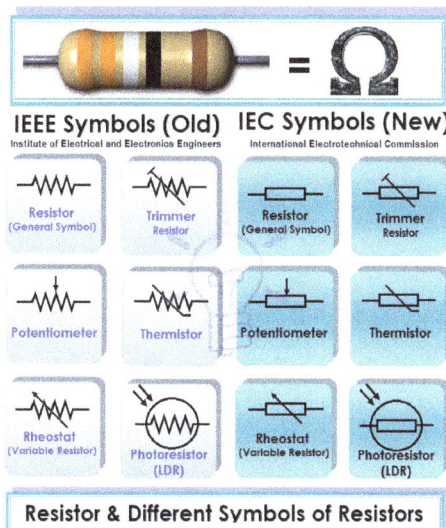

Resistors and Symbols of Different Types of Resistors. IEEE & IEC symbols of Resistors

Types of Resistors

Resistors are available in different size, Shapes and materials. We will discuss all possible resistor types one by one in detail with pro and cons and application/uses.

Different Types of Resistor

There are two basic types of resistors:

1. Linear Resistors

2. Non Linear Resistors

Linear Resistors

Those resistors, which values change with the applied voltage and temperature, are called linear resistors. In other words, a resistor, which current value is directly proportional to the applied voltage is known as linear resistors.

Generally, there are two types of resistors which have linear properties:

• Variable Resistors

• Fixed Resistors

As the name tells everything, fixed resistor is a resistor which has a specific value and we can't change the value of fixed resistors.

Types of Fixed Resistors

1) Carbon Composition Resistors

2) Wire Wound Resistors

3) Thin Film Resistors

4) Thick Film Resistors

Carbon Composition Resistors

A typical fixed resistor is made from the mixture of granulated or powdered carbon or graphite, insulation filler, or a resin binder. The ratio of the insulation material determines the actual resistance of the resistor. The insulating powder (binder) made in the shape of rods and there are two metal caps on the both ends of the rod.

There are two conductor wires on the both ends of the resistor for easy connectivity in the circuit via soldering. A plastic coat covers the rods with different color codes (printed) which denote the resistance value. They are available in 1 ohm to 25 mega ohms and in power rating from ¼ watt to up to 5 Watts.

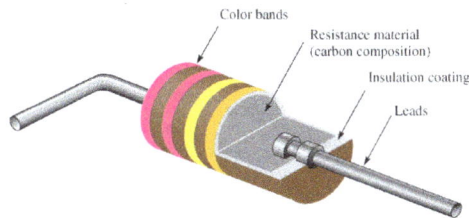

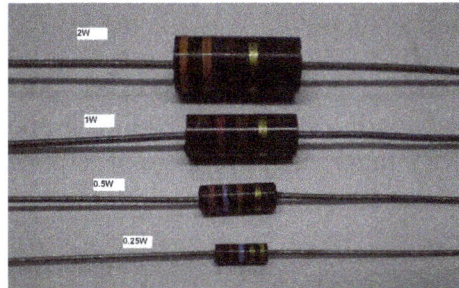

Carbon Composition Resistors. Construction and Wattage Rating

Characteristic of Fixed Resistors

Generally, they are very cheap and small in size, hence, occupy less space. They are reliable and available in different ohmic and power ratings. Also, fixed resistor can be easily connected to the circuit and withstand for more voltage.

In other hand, they are less stable means their temperature coefficient is very high. Also, they make a slight noise as compared to other types of resistors.

Wire Wound Resistors

Wire wound resistor is made from the insulating core or rod by wrapping around a resistive wire. The resistance wire is generally Tungsten, manganin, Nichrome or nickel or nickel chromium alloy and the insulating core is made of porcelain, Bakelite, press bond paper or ceramic clay material.

The manganin wire wound resistors are very costly and used with the sensitive test equipments e.g. Wheatstone bridge, etc. They are available in the range of 2 watts up to 100 watt power rating or more. The ohmic value of these types of resistors is 1 ohm up to 200k ohms or more and can be operated safely up to 350°C.

in addition, the power rating of a high power wire wound resistor is 500 Watts and the available resistance value of these resistors are is 0.1 ohm – 100k Ohms.

Wire wound Resistors

Advantages and Disadvantage of Wire wound Resistors

Wire wound resistors make lower noise than carbon composition resistors. Their performance is well in overload conditions. They are reliable and flexible and can be used with DC and Audio frequency range. Disadvantage of wire wound resistor is that they are costly and can't be used in high frequency equipments.

Application/uses of Wire Wound Resistors

Wire wound resistors used where high sensitivity, accurate measurement and balanced current control is required, e.g. as a shunt with ampere meter. Moreover, Wire wound resistors are generally used in high power rating devices and equipments, Testing and measuring devices, industries, and control equipments.

Thin Film Resistors

Basically, all thin film resistors are made of from high grid ceramic rod and a resistive material. A very thin conducting material layer overlaid on insulating rod, plate or tube which is made from high quality ceramic material or glass. There are two further types of thin film resistors:

1.   Carbon Film Resistors

2.   Metal Film Resistors

Carbon Film Resistors

Carbon Film resistors contains on an insulating material rod or core made of high grade ceramic material which is called the substrate. A very thin resistive carbon layer or film overlaid around the rod. These kinds of resistors are widely used in electronic circuits because of negligible noise and wide operating range and the stability as compared to solid carbon resistors.

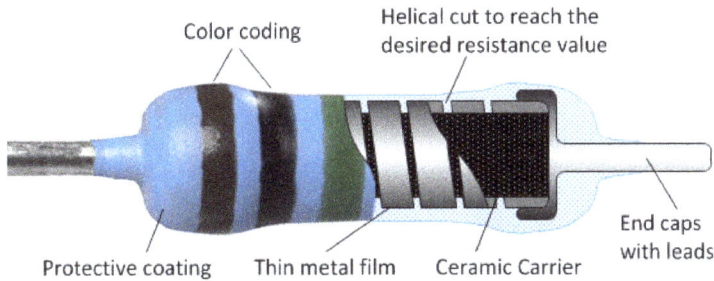

Color coding   Helical cut to reach the desired resistance value

Protective coating   Thin metal film   Ceramic Carrier   End caps with leads

Construction of Carbon Film Resistors & Its labels

Metal Film Resistors

Metal film resistors are same in construction like Carbon film resistors, but the main difference is that there is metal (or a mixture of the metal oxides, Nickel Chromium or mixture of metals and glass which is called metal glaze which is used as resistive film) instead of carbon. Metal film resistors are very tiny, cheap and reliable in operation. Their temperature coefficient is very low (± 2 ppm/°C) and used where stability and low noise level is important.

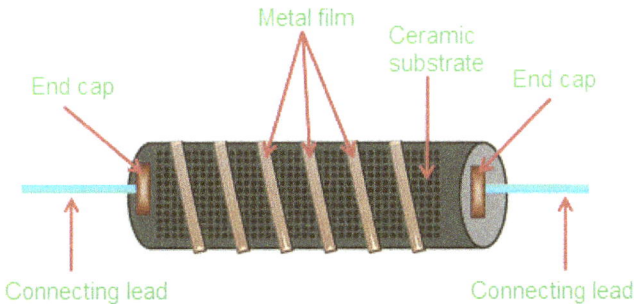

Metal film   Ceramic substrate

End cap   End cap

Connecting lead   Connecting lead

Metal Film Resistor, Construction and name of internal parts

Thick Film Resistors

The production method of Thick film resistors is same like thin film resistors, but the difference is that there is a thick film instead of a thin film or layer of resistive material around. That's why it is called Thick film resistors. There are two additional types of thick film resistors:

1. Metal Oxide Resistors

2. Cermet Film Resistors

3. Fusible Resistors

Metal Oxide Resistors

By oxidizing a thick film of Tin Chloride on a heated glass rod (substrate) is the simple method to make a Metal oxide Resistor. These resistors are available in a wide range of resistance with high temperature stability. In addition, the level of operating noise is very low and can be used at high voltages.

Cermet Oxide Resistors

In the cermet oxide resistors, the internal area contains on ceramic insulation materials. And then a carbon or metal alloy film or layer wrapped around the resistor and then fix it in a ceramic metal (which is known as Cermet). They are made in the square or rectangular shape and leads and pins are under the resistors for easy installation in printed circuit boards. They provide a stable operation in high temperature because their values do not change with change in temperature.

Cermet Film Resistor

Fusible Resistors

These kinds of resistors are same like a wire wound resistor. When a circuit power rating increased than the specified value, then this resistor is fused, i.e. it breaks or open the circuit. That's why it is called Fusible resistors. Fusible restores perform double jobs means they limit the current as well as it can be used as a fuse.

They used widely in TV Sets, Amplifiers, and other expensive electronic circuits. Generally, the ohmic value of fusible resistors is less than 10 Ohms.

Variable Resistors

As the name indicates, those resistors which values can be changed through a dial, knob, and screw or manually by a proper method. In these types of resistors, there is a sliding arm, which is connected to the shaft and the value of resistance can be changed by rotating the arm. They are used in the radio receiver for volume control and tone control resistance.

Following are the further types of Variable Resistors:

1. Potentiometers

2. Rheostats

3. Trimmers

Potentiometers

Potentiometer is a three terminal device which is used for controlling the level of voltage in the circuit. The resistance between two external terminals is constant while the third terminal is connected with moving contact (Wiper) which is variable. The value of resistance can be changed by rotating the wiper which is connected to the control shaft.

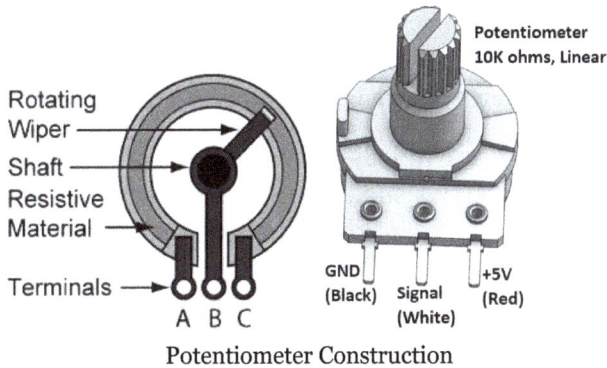

Potentiometer Construction

This way, Potentiometers can be used as a voltage divider and these resistors are called variable composition resistors. They are available up to 10 Mega Ohms.

Different Types of Potentiometers

Rheostats

Rheostats are a two or three terminal device which is used for the current limiting purpose by hand or manual operation. Rheostats are also known as tapped resistors or variable wire wound resistors.

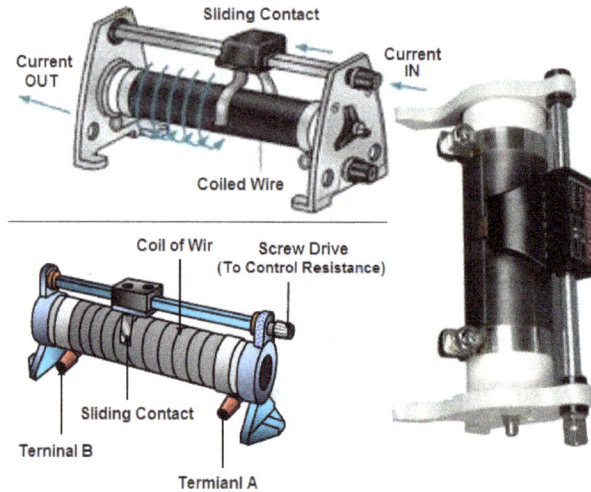

Types of Rheostats resistor and construction of Screw Drive Rheostat

To make a rheostat, they wire wind the Nichrome resistance around a ceramic core and then assembled in a protective shell. A metal band is wrapped around the resistor element and it can be used as a Potentiometer or Rheostats.

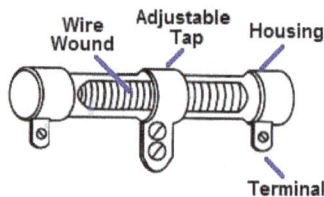

Construction of Tapped Rheostat

Variable wire wound resistors are available in the range of 1 ohm up to 150 Ohms. The available power rating of these resistors is 3 to 200 Watts. While the most used Rheostats according to power rating is between 5 to 50 Watts.

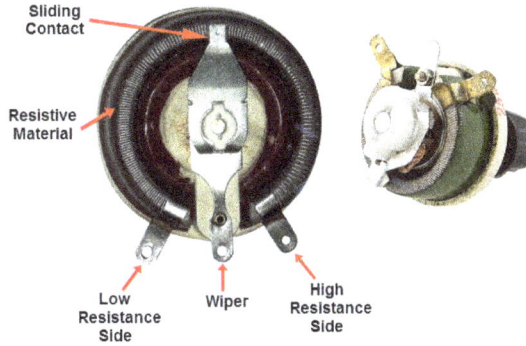

Sliding Contact
Resistive Material
Low Resistance Side
Wiper
High Resistance Side

DDifference between Potentiometer and Rheostats

Basically, there is no difference between Potentiometer and Rheostat. Both are variable resistors. The main difference is the use and circuit operation, i.e. for which purpose we use that variable resistor.

For example, if we connect a circuit between resistor element terminals (where one terminal is a general end of the resistor element while the other one is sliding contact or wiper) as a variable resistor for controlling the circuit current, then it is Rheostats.

In the other hand, if we do the same as mentioned above for controlling the level of voltage, then this variable resistor would be called a potentiometer. That's it.

Trimmers

There is an additional screw with Potentiometer or variable resistors for better efficiency and operation and they are known as Trimmers. The value of resistance can be changed by changing the position of screw to rotate by a small screwdriver.

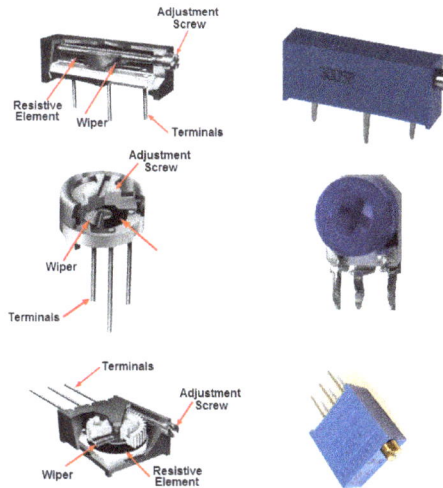

Adjustment Screw
Resistive Element
Wiper
Terminals
Adjustment Screw
Wiper
Terminals
Terminals
Adjustment Screw
Wiper
Resistive Element

Construction of Different Types of Trimmer.
Trimmer potentiometer Resistor construction

They are made from carbon composition, carbon film, cermet and wire materials and available in the range of 50 Ohms up to 5 mega ohms. The power rating of Trimmers potentiometers are from 1/3 to ¾ Watts.

Non Linear Resistors

We know that, nonlinear resistors are those resistors, where the current flowing through it does not change according to Ohm's Law but, changes with change in temperature or applied voltage.

In addition, if the flowing current through a resistor changes with change in body temperature, then these kinds of resistors are called Thermistors. If the flowing current through a resistor change with the applied voltages, then it is called a Varistors or VDR (Voltage Dependent Resistors).

Following are the additional types of Non Linear Resistors:

1. Thermisters

2. Varisters (VDR)

3. Photo Resistor or Photo Conductive Cell or LDR

Thermisters

Thermisters is a two terminal device which is very sensitive to temperature. In other words, Thermisters is a type of variable resistor which notices the change in temperature. Thermisters are made from the cobalt, Nickel, Strontium and the metal oxides of Manganese. The Resistance of a Thermister is inversely proportional to the temperature, i.e. resistance increases when temperature decrease and vice versa.

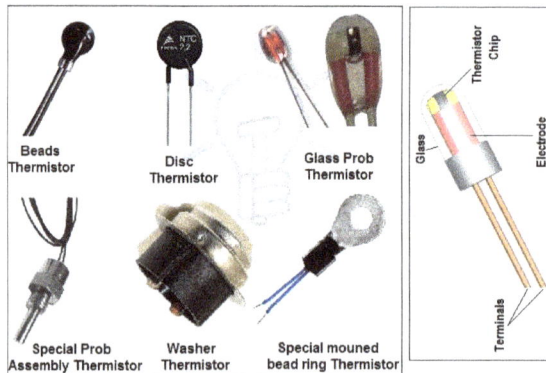

Types of Thermisters & Its Construction

It means, Thermisteres has a negative temperature coefficient (NTC) but there is also a PTC (Positive Temperature Coefficient) which a made from pid barium titanate semiconductor materials and their resistance increases when increases in temperature.

Varisters (VDR)

Varisters are voltage dependent Resistors (VDR) which is used to eliminate the high voltage transients. In other words, a special type of variable resistors used to protect circuits from destructive voltage spikes is called varisters.

When voltage increases (due to lighting or line faults) across a connected sensitive device or system, then it reduces the level of voltage to a secure level i.e. it changes the level of voltages.

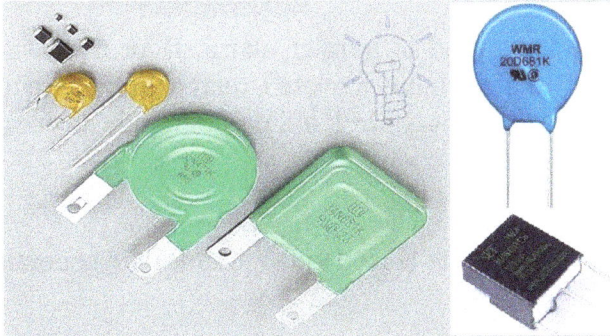

Types of Varisters

Photo Resistor or Photo Conductive Cell or LDR

Photo Resistor or LDR (Light Dependent Resistors) is a resistor which terminal value of resistance changes with light intensity. In other words, those resistors, which resistance values changes with the falling light on their surface is called Photo Resistor or Photo Conductive Cell or LDR (Light Dependent Resistor). The material which is used to make these kinds of resistors is called photo conductors, e.g. cadmium sulphide, lead sulphide etc.

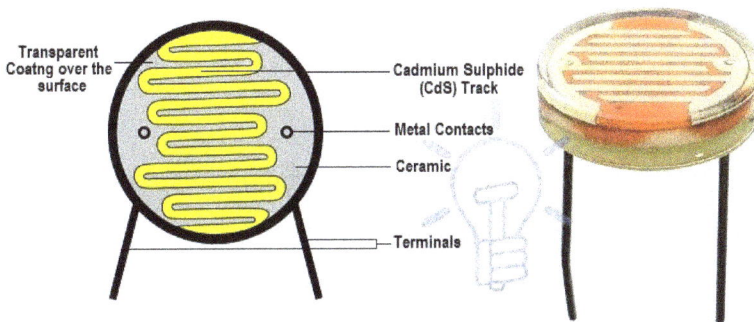

Construction of LDR (Light Dependent Resistor), Photo-resistor or photo conductive cell

When light falls on the photoconductive cells (LDR or Photo resistor), then there is an increase in the free carriers (electron hole pairs) due to light energy, which reduce the resistance of semiconductor material (i.e. the quantity of light energy is inversely proportional to the semiconductor material). It means photo resistors have a negative temperature coefficient.

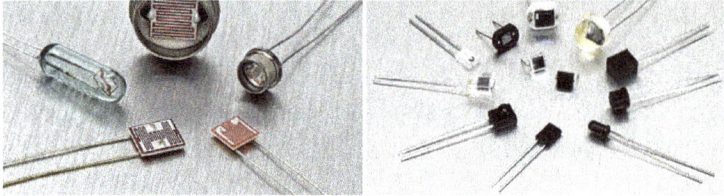
Types of Photo cells, and LDR

Application and uses of Photo Resistors/Photo Conductive Cells or LDR

These types of resistors are used in burglar alarm, Door Openers, Flame detectors, Smock detectors, light meters, light activated relay control circuits, industrial, and commercial automatic street light control and photographic devices and equipments.

Uses/Application of Resistors

Practically, both types of resistors (Fixed and Variable) are generally used for the following purposes.

Resistors are used:

- For Current control and limiting
- To change electrical energy in the form of heat energy
- As a shunt in Ampere meters
- As a multiplier in a Voltmeter
- To control temperature
- To control voltage or Drop
- For protection purposes, e.g. Fusible Resistors
- In laboratories
- In home electrical appliances like heater, iron, immersion rod etc.
- Widely used in the electronics industries

Capacitors

A capacitor is a passive two terminal component which stores electric charge. This component consists of two conductors which are separated by a dielectric medium. The potential difference when applied across the conductors polarizes the dipole ions to store the charge in the dielectric medium. The circuit symbol of a capacitor is shown below:

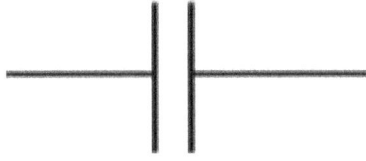

The capacitance or the potential storage by the capacitor is measured in Farads which is symbolized as 'F'. One Farad is the capacitance when one coulomb of electric charge is stored in the conductor on the application of one volt potential difference.

The charge stored in a capacitor is given by,

$Q = CV$

Where, Q - charge stored by the capacitor

C - Capacitance value of the capacitor

V - Voltage applied across the capacitor

Note the other formula of current, $I = dQ/dt$.

Taking the derivative with respect to time,

$dQ/dt = d(CV)/dt$

From the above statement, we can express the equation as:

$I = C\,(dV/dt)$

As you turn on the power supply, the current begins to flow through the capacitor inducing the positive and negative potentials across its plates. The capacitor continues to charge until the capacitor voltage equalizes up to the supply voltage which is called as the charging phase of the capacitor. Once the capacitor is fully charged at the end of this phase, it gets open circuited for DC. It begins to discharge when the power of the capacitor is switched off. The charging and discharging of the capacitor is given by a time constant.

$\tau = RC$

The voltage across the capacitor is given by,

$$V_C = V^*\left(1 - e^{-t/\tau}\right)\left[\text{for charging}\right]$$
$$V_C = V^* e^{-t/\tau}\left[\text{for discharging}\right]$$

Capacitors are widely used in a variety of applications of electronic circuits such as:

- Store charges such as in a camera flash circuit.

- Smoothing the output of power supply circuits.

- Coupling of two stages of a circuit (coupling of an audio stage with a loud speaker).

- Filter networks(tone control of an audio system).

- Delay applications (as in 555 timer IC controlling the charging and discharging)

- Tuning radios to particular frequencies.

- Phase alteration.

The conductors offer a series resistance and if the capacitor is constructed using tubular structure then some inductance is also induced. The dielectric medium between the plates has an electric field strength limit and also passes a small amount of leakage current which results into a Breakdown voltage.

There are different types of capacitors; they can be fixed or variable. They are categorized into two groups, polarized or non-polarized. Electrolytic capacitors are polarized. Most of the low value capacitors are non-polarized. The symbol of capacitors from each group is shown below:

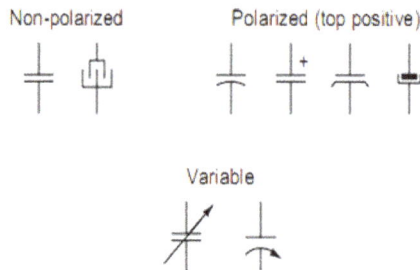

Construction and Types:

The capacitor consists of two conducting plates that are separated by an insulating medium known as the dielectric. The capacitance is dependent upon the surface area of the plates, the distance between the dielectric medium and the dielectric constant of the object. The greater the area of the plates, the closer they are together and greater the value of the dielectric constant the greater is the value of capacitance. High capacitance capacitors are now available in small size. This has been achieved employing a number of techniques like having several sets of plates, placing the plates very close to one another, having a thin layer of dielectric placed between them and developing special insulating dielectric materials.

The capacitance of a capacitor is also affected by the shape or structure of the capacitors. The capacitors are available in different shapes like radial lead type which are rectangular or cubical or axial lead type which are tubular or cylindrical.

The variable type of capacitors can vary the capacitance by changing the distance between the plates or the effective area of the capacitor.

The polarized type of capacitors should be connected as per their polarity or else the capacitor can be damaged due to incorrect connection.

The low value capacitors are non-polarized and can be connected in any manner. They are not damaged by heat when soldering, except for the polystyrene type of capacitor. They have high voltage ratings of at least 50V, usually 250V or so many small value capacitors have their value printed but without a multiplier, so you need to use experience to work out what the multiplier should be.

For example:

- 0.1 means 0.1μF = 100nF.

Sometimes the multiplier is used in place of the decimal point:

For example - 4n7 means 4.7nF.

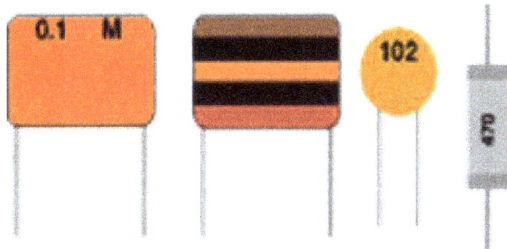

The various types of capacitors are given below:

Fixed Capacitors:

- Film Capacitors like glass capacitor, mica capacitors, silver mica capacitor, ceramic capacitor, paper capacitor, metalized paper capacitor, polyester capacitor, polystyrene capacitor, metalized polyester capacitor, polycarbonate capacitor, polypropylene capacitors, Teflon capacitors, porcelain capacitor.

- Electrolyte Capacitors like aluminum electrolyte, tantalum electrolyte, aluminum-tantalum electrolyte.

Types of Fixed Capacitors

Film Capacitors

Film capacitors consist of a relatively large family of capacitors with the difference being in their dielectric properties. These include polyester (Mylar), polystyrene, polypropylene, polycarbonate, metalized paper, Teflon etc. Film type capacitors are available in capacitance ranges from 5pF to 100uF depending upon the actual type of capacitor and its voltage rating. Film capacitors come in various shapes and case styles such as:

- Wrap and Fill (oval & round): The capacitor is wrapped in a tight plastic tape and the ends are filled with epoxy in order to seal them.

- Epoxy Case (rectangular & round): The capacitor is encased in a molded plastic shell which is filled with epoxy.

- Metal hermetically sealed (rectangular & round): The capacitor is encased in a metal tube or a can and sealed with epoxy.

All the above case styles are available in both Axial and Radial leads.

Paper Capacitor

Paper capacitors are made of paper or oil-impregnated paper and aluminum foil layers rolled into a cylinder and sealed with wax. These capacitors were commonly used but are now replaced by the plastic or polymer type of capacitors. The paper capacitors are bulky, highly hygroscopic and soaks moisture which causes loss to the dielectric degrading its overall performance is the major drawback with this type of capacitors. The other variants include oil-impregnated, paper-polyester and Kraft paper capacitor.

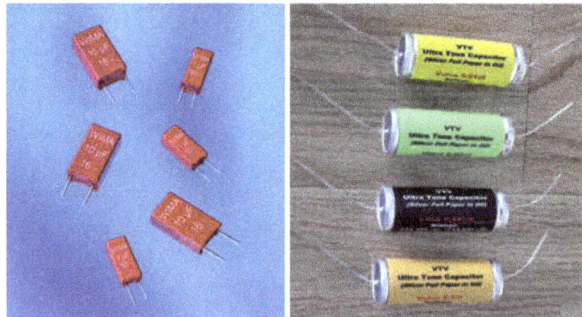

Paper capacitor   Oil-Impregnated paper capacitor

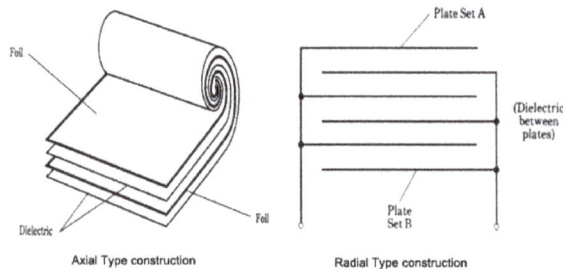

Axial Type construction   Radial Type construction

Metalized Paper Capacitors

The metalized paper capacitors are smaller in size than the conventional paper capacitors. However, these capacitors are appropriate for only low current applications and are now replaced by metalized film capacitors.

Axial type metalized paper   Radial type metalized paper

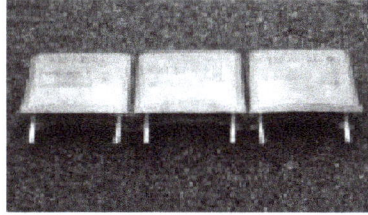

Mica Capacitor

The mica capacitor uses mica as the dielectric medium. Mica is inert in nature and hence the physical and chemical properties do not change as it ages. It provides good temperature stability and resistance to corona discharge i.e. electrical discharges due to ionization around conductor. However, the cost is very high and due to improper sealing the capacitor is highly prone to moisture which increases the power factor.

Construction of Mica capacitor   Original photographs of Mica capacitors

Silver Mica or Metalized Mica Capacitor

These are a kind of mica capacitor which has an additional advantage of reduced moisture infiltration. These capacitors are expensive and are used often in HF and low VHF radio frequency circuits as low value accurate capacitors particularly in the oscillators and filters. The reasons that these capacitors are still in use regardless of high cost, large size and availability of other low cost capacitors are due to its remarkable features such as:

- Low tolerance of +/- 1%.

- Positive temperature coefficient of 35 to 75 ppm/C.

- Greater range from few pF to two or three pF.

- Little voltage dependence.

- High stability.

- Good Q factor.

However, these capacitors are not widely used these days.

Silver mica or metalized mica capacitor

Glass Capacitor

These capacitors are fabricated of glass dielectrics and are very expensive which are used for highly accurate, stable and reliable operation in harsh environmental conditions. These are resistant to nuclear radiations and available in range of 10pF to 1000pF.

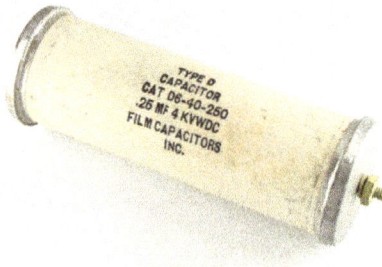

Glass Capacitor

Ceramic Capacitor

The non – polarized type ceramic capacitors which are also known as 'Disc capacitors' are widely used these days. These are available in millions of varieties of cost and performance. The features of ceramic capacitor depend upon:

- Type of ceramic dielectric used in the capacitor which varies in the temperature coefficient.

- Dielectric losses.

The exact formulas of the different ceramics used in ceramic capacitors vary from one manufacturer to another. The common compounds such as titanium dioxide, strontium titanate, and barium titanate are the three main types available although other types such as leaded disc ceramic capacitors for through hole mounting which are resin coated, multilayer surface mount chip ceramic capacitors and microwave bare leadless disc ceramic capacitors that are designed to sit in a slot in the PCB and are soldered in place.

These are made by placing silver coated ceramic plates on two sides and assembled together to form the capacitor. The surface mount version consists of the ceramic dielectric in which a number of interleaved precious metal electrodes are contained. This structure gives rise to a high capacitance per unit volume. The inner electrodes are connected to the two terminations, either by silver palladium (AgPd) alloy in the ratio 65 : 35, or silver dipped with a barrier layer of plated nickel and finally covered with a layer of plated tin (NiSn).

The Electronics industries alliance (EIA) has broadly classified the ceramics used in these capacitors into 3 classes – class 1, class 2 and class 3. The lower is the class better are its overall characteristics but is on the cost of size. Each class defines the working temperature range, temperature drift, tolerance, etc. The typical values range from 10pF to 1uF. The capacitance values are labeled by three digit codes where the first two digits represent a number and the third digit is the multiplier digit.

For example: 103 means $10 * 10^3$ pF which is 0.01uF

or

104 which is $10*10^4$ pF which is 0.1uF

The tolerance is indicated by a letter like j=5%, K=10% and M=20%.

These capacitors are commonly used as a timing element in filter circuit and balancing oscillator circuits in radio frequency applications, coupling and decoupling networks.

The three ceramic classes decided by EIA are:

Class1 - Class 1 ceramic capacitors are the most stable forms of ceramic capacitor with respect to temperature. The common compounds used as the dielectrics are magnesium titanate for a positive temperature coefficient (PTC), or calcium titanate for capacitors with a negative temperature coefficient (NTC). Using combinations of these and other compounds it is possible to obtain a dielectric constant of between 5 and 150. They have an almost linear characteristic and their properties are almost independent of frequency within normal bounds. Temperature coefficients between +40 and -5000 ppm/C can be obtained.

Class 1 capacitors offer the best performance with respect to dissipation factor. A typical figure may be 0.15%. It is also possible to obtain very high accuracy (~1%) class 1 capacitors rather than the more usual 5% or 10% tolerance versions. The highest accuracy class 1 capacitors are designated C0G or NP0.

EIA has defined a set of codes in order to have a managed way of ceramic capacitor performance. The codes of class 1 and class 2 capacitors are different.

The class 1 codes are as follows:

First Character (Letter) Significant Figures		Second Character (Digit) Multiplier		Third Character (Letter) Tolerance	
C	0.0	0	-1	G	+/-30
B	0.3	1	-10	H	+/-60
L	0.8	2	-100	J	+/-120
A	0.9	3	-1000	K	+/-250
M	1.0	4	+1	L	+/-500
P	1.5	5	+10	M	+/-1000
R	2.2	6	+100	N	+/-2500
S	3.3	7	+1000		
T	4.7	8			
V	5.6				
U	7.5				

The first character is a letter which gives the significant figure of the change in capacitance over temperature in ppm/C.

- The second character is numeric and gives the multiplier.

- The third character is a letter and gives the maximum error in ppm/C.

One common example of class 1 capacitor is a CoG. This has 0 drift, with an error of 30PPM/C.

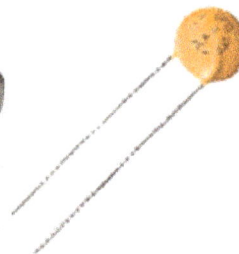

COG Ceramic   NPO Ceramic

Class 2 - Class 2 capacitors are better in size, but have less accuracy and stability. As a result, they are normally used for decoupling, coupling and bypass applications where accuracy is not of prime importance. A typical class 2 capacitor may change capacitance by 15% or so over a -50C to +85C temperature range and it may have a dissipation factor of 2.5%. It will have average to poor accuracy (from 10% down to +20/-80%). However for many applications these figures would not present a problem.

The class 2 codes are as follows:

First Character (Letter) Low Temperature		Second Character (Digit) High Temperature		Third Character (Letter) Change	
X	-5SC(-67F)	2	+45C(+113F)	D	+/-3.3%
Y	-30C(-22F)	4	+65(+149F)	E	+/-4.7%
Z	+10C(+50F)	5	+85(+185F)	F	+/-7.5%
		6	+105(+221F)	P	+/-10%
		7	+125(+257F)	R	+/-15%
				S	+/-22%
				T	+22%/-33%
				U	+22%/-56%
				V	+22%/-82%

The first character is a letter which gives the low-end operating temperature.

- The second is numeric which provides the high-end operating temperature.

- The third character is a letter which provides capacitance change over that temperature range.

The common examples of class 2 ceramic capacitors are:

- The X7R capacitor which operates from -55C to +125C with a capacitance change of up to 15%.

- The Z5U capacitor which operates from +10C to +85C with a capacitance change of up to +22% to -56%.

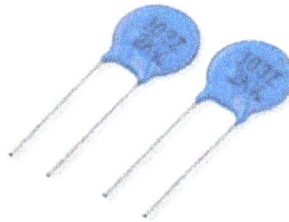

X7R ceramic   Z5U ceramic

Class 3 - Class 3 ceramic capacitors are small in size with less accuracy, stability and low dissipation factor. This type of capacitors cannot withstand high voltages.

Barium titanate that has a dielectric constant about 1250 is used as the dielectric. A typical class 3 capacitor will change its capacitance by -22% to +50% over a temperature range of +10C to +55C. It may also have a dissipation factor of around 3 to 5%. It will have a fairly poor accuracy (commonly, 20%, or -20/+80%). Therefore, class 3 ceramic capacitors are typically used as decoupling or in other power supply applications where accuracy is not of prime importance. However, they must not be used in applications where spikes are present as they cannot withstand high voltages.

SMT ceramic capacitors are also available in standard packages which have following designations given in the below table:

Package Designation	Size (Mm)	Size (Inches)
1812	4.6 3.0	0.18 0.12
1206	3.0 1.5	0.12 0.06
0805	2.0 1.3	0.08 0.05)
0603	1.5 0.8	0.06 0.03
0402	1.0 0.5	0.04 0.02
0201	0.6 0.3	0.02 0.01

Plastic Capacitors

i. Polyester or PET capacitor:

Polyester or PET capacitors are plastic capacitors available as leaded packages that replace the paper capacitors. These capacitors are made of polyester films which small in size and available at low cost. These have Operating voltages up to 60,000 V DC, operating temperatures up to 125° C and low moisture absorption. These are mostly used as low frequency signal capacitors and integrators. They are preferred where cost plays an important role because they have high tolerances of 5 - 10 %.

ii. Polystyrene capacitors:

These are large size capacitors available in tubular shape leaded packages. They have high stability, negative temperature coefficient (NTC), high accuracy and low moisture absorption. The operating temperature is limited to +85 C. These are mostly preferred for low frequency applications as the tubular structure induces inductances which degraded the performance at high frequencies.

iii. Kapton polyimide capacitor:

These capacitors are similar to polyester or PET capacitors that are made of Kapton polyimide film. They are expensive but offer high operating temperatures up to 25° C. These capacitors are not suitable for RF applications.

iv. Polycarbonate capacitors:

These are high performance capacitors which are least affected as it ages. These are characterized by good insulation resistance and dissipation factor. The operating temperature ranges from -55 to +125 C. The dielectric constant is 3.2 %, and dielectric strength is 38 KV/mm. The dissipative factor is 0.0007 at 50Hz and 0.001 at 1MHz. The water absorption is 0.16%. These are mostly used for filters, coupling and timing applications. These can be directly replaced by Polyethylene napthalate (PEN), Polyphenylene sulphide (PPS), Polyimide (PI) and Polytetrafluoroethylene (PTFE).

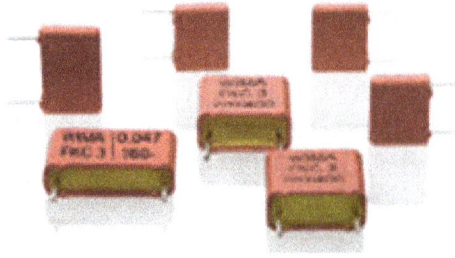

v. Polypropylene capacitors:

These are used where higher tolerances than PET film capacitors are demanded. These are available in leaded packages and are used for low frequency operation. They have high operating voltages and are resistant to the breakdown. However, they get damaged by transient over-voltages or voltage reversals.

vi. Polysulfone capacitor:

These capacitors are similar to polycarbonate capacitors but can withstand full voltage at comparatively higher temperatures. These capacitors are very expensive and are not readily available. The stability is limited as the moisture absorption is typically 0.2%.

vii. TEFLON or PTFE fluorocarbon capacitor:

These plastic capacitors are large in size and expensive. Due to low losses and higher stability these are used for some critical applications. The operating temperature ranges up to 25° C. The dielectric used is Polytetrafluoroethylene.

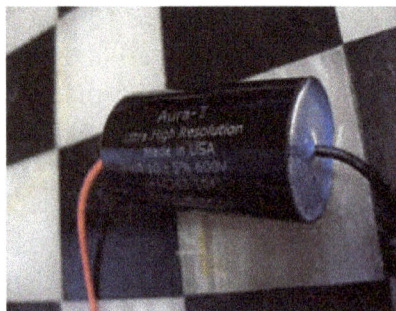

TEFLON capacitor

Viii. Polyamide capacitor:

These plastic film capacitors are large in size and expensive. The operating temperature ranges up to 20° C.

ix. Metalized polyester or Metalized plastic capacitor:

These capacitors have metalized plastic films which provide self-heating advantage and also reduce the size of the capacitor over conventional plastic or polyester capacitor. However, they are limited by maximum current capacity. They are available in leaded package.

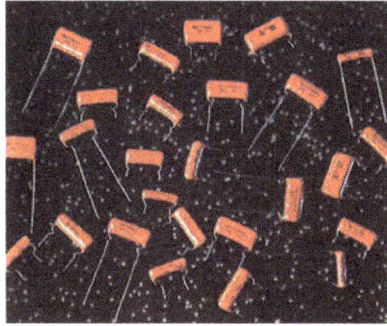
Metalized polyester capacitors

Electrolyte Capacitors

i. Aluminum Electrolyte Capacitor

These polarized capacitors are made of oxide film on aluminum foils. These are cheaper and easily available. The range of values typically varies from 1uF to 47000uF and large tolerance of 20%. The voltage ratings range up to 500V. They have high capacitance to volume ratio and used for smoothing in power supply circuits or coupling capacitors in audio amplifiers. These are available in both leaded and surface mount packages. The capacitance value and voltage ratings are either printed in uF or coded by a letter followed by three digits. The three digits represent the capacitance value in pF where first two digits represent the number and the third one is the multiplier digit. The letter codes are as follows:

Letter	Voltage
e	2.5
G	4
J	6.3
A	10
C	16
D	20
E	25
V	35
H	50

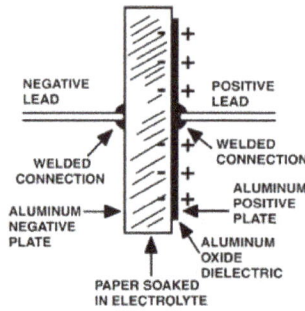

Aluminum electrolyte construction

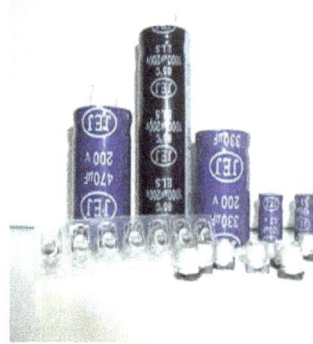

Aluminum electrolyte capacitor

ii. Tantalum Electrolyte Capacitor

These capacitors utilize tantalum oxide which enables the fabrication of small size electrolytes. These are costlier than aluminum electrolytes and have lower maximum voltage up to 50V and are preferred where size matters. Their typical values range from 47uF to 470uF. These may be using tantalum oxide layered foils or porous anode with sulphuric acid as electrolyte in between tantalum foils in a wet tantalum electrolyte or solid tantalum electrolytes. Their SMT formats are available in standard packages where package designations have been defined by the EIA.

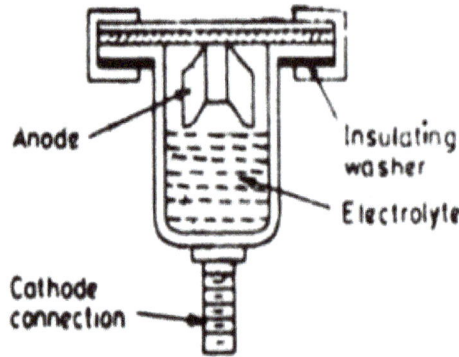

A liquid tantalum electrolyte

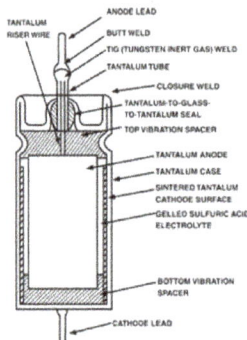

Tantalum Electrolyte Construction

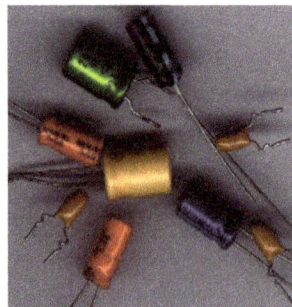

Tantalum Electrolyte Capacitors

iii. Super Capacitor

Super capacitors also called as electrolyte double layer capacitor are made of a thin electrolyte separator flanked with activated carbon ions. These have capacitance values as high as of order of mille farads. These are used as temporary power source as a replacement of batteries.

Super capacitors

Variable Capacitors

The variable type of capacitors can vary the capacitance by changing the distance between the plates or the effective area of the capacitor.

Types of Variable Capacitor

Air-gap Capacitors

These capacitors use air as the dielectric medium. The distance between the plates can be varied to change the capacitance. The capacitance values offered are high and can be used with high voltages. These are used for high frequency operations in communication systems.

Vacuum Capacitors

These capacitors have glass or ceramic encapsulation and vacuum as the dielectric. Their complex construction makes it very expensive. Theoretically, it has less losses and are used in RF applications.

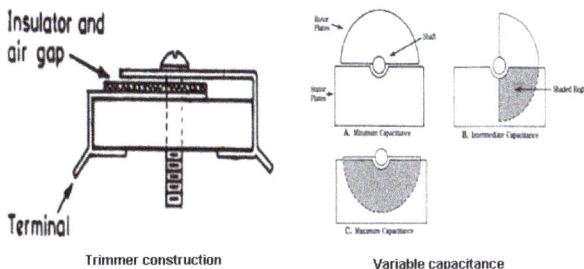

Trimmer construction

Variable capacitance

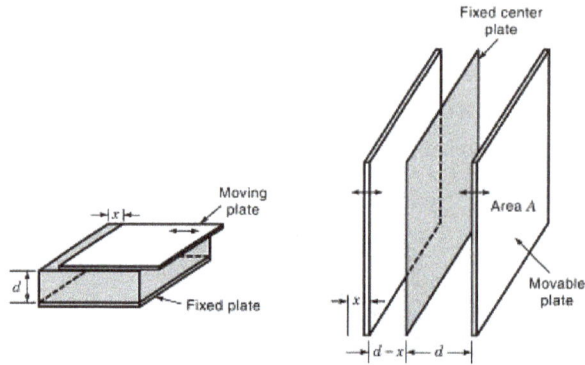

Principle behind variable capacitors

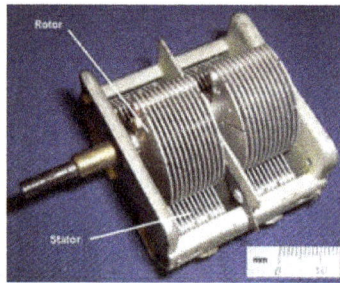

Air-gap variable   Trimmer

Capacitor Color Code

Color Code	
Color	Number
Black	0
Brown	1
Red	2
Orange	3
Yellow	4
Green	5
Blue	6
Violet	7
Grey	8
White	9

color code was used on polyester capacitors for many years. It is now obsolete, but of course there are many still around. The colors should be read like the resistor code:

- The top three color bands give the value in pF.

- The 4th band is for tolerance.

- The 5th band is for the voltage rating.

For example,

Brown, black, orange   means 10000pF = 10nF = 0.01μF.

There are no gaps between the colour bands, so 2 identical bands actually appear as a wide band.

Wide red, yellow means 220nF = 0.22µF.

Inductors

The inductor is a passive component which stores the electrical energy in the magnetic field when the electric current passes through it. Or we can say that the inductor is an electrical device which possesses the inductance.

The inductor is made of wire which has the property of inductance, i.e., opposes the flow of current. The inductance of wire increases by increasing the number of turns. The alphabet 'L' is used for representing the inductor, and it is measured in Henry. The inductance characterises the inductor. The figure below shows the symbolic representation of inductor.

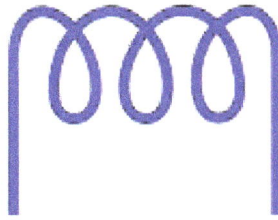

The electric current I flow through the coil generates the magnetic field around it. Consider the magnetic field generates the flux Φ when current flows through it. The ratio of the flux and the current gives inductances,

$$L = \frac{\theta}{I}$$

The inductance of the circuit depends on the current paths and the magnetic permeability of the nearer material. The magnetic permeability shows the ability of the material to forms the magnetic field.

There are different types of inductors. Depending on their material type they are basically categorized as follows:

Air Core Inductor

Ceramic core inductors are referred as "Air core inductors". Ceramic is the most commonly used material for inductor cores. Ceramic has very low thermal co-efficient of expansion, so even for a range of operating temperatures the stability of the inductor's inductance is high. Since ceramic has no magnetic properties, there is no increase in the permeability value due to the core material.

Air Core Inductor

Its main aim is to give a form for the coil. In some cases it will also provide the structure to hold the terminals in place. The main advantage of these inductors are very low core losses, high Quality factor. These are mainly used in high frequency applications where low inductance values are required.

Iron Core Inductor

In the areas where low space inductors are in need then these iron core inductors are best option. These inductors have high power and high inductance value but limited in high frequency capacity. These are applicable in audio equipments. When compared with other core indictors these have very limited applications.

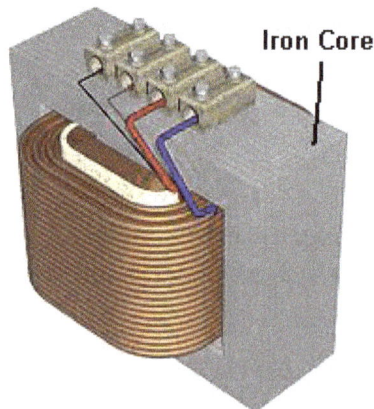

Inductor with iron core

Ferrite Core Inductor

Ferrite is also referred as ferromagnetic material. They exhibit magnetic properties. They consist of mixed metal oxide of iron and other elements to form crystalline structures. The general composition of ferrites is XFe_2O_4. Where X represents transition materials. Mostly easily magnetized material combinations are used such as manganese and zinc (MnZn), nickel and zinc (niZn).

Ferrites are mainly two types they are soft ferrites and hard ferrites. These are classified according to the magnetic coercivity. Coercivity is the magnetic field intensity needed to demagnetize the ferromagnetic material from complete saturation state to zero.

Soft Ferrite

These materials will have the ability to reverse their polarity of their magnetization without any particular amount of energy needed to reverse the magnetic polarity.

Hard Ferrite

These are also called as permanent magnets. These will keep the polarity of the magnetization even after removing the magnetic field.

Ferrite core inductor will help to improve the performance of the inductor by increasing the permeability of the coil which leads to increase the value of the inductance. The level of the permeability of the ferrite core used within the inductors will depend on the ferrite material. This permeability level ranges from 20 to 15,000 according to the material of ferrite. Thus the inductance is very high with ferrite core when compared to the inductor with air core.

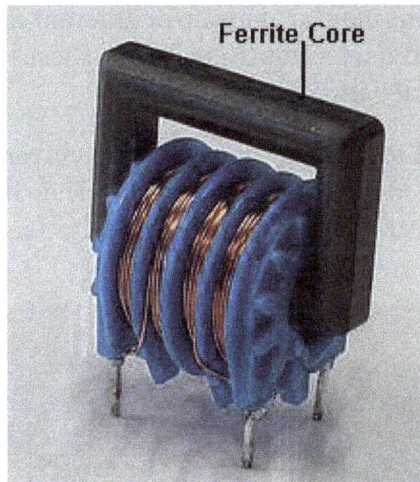

Inductor with ferrite core

Iron Powder Inductor

These are formed from very fine particles with insulated particles of highly pure iron powder. This type of inductor contains nearly 100% iron only. It gives us a solid looking core when this iron power is compressed under very high pressure and mixed with a binder such as epoxy or phenolic. By this action iron powder forms like a magnetic solid structure which consists of distributed air gap.

Due to this air gap it is capable to store high magnetic flux when compared with the ferrite core. This characteristic allows a higher DC current level to flow through the inductor before inductor saturates. This leads to reduce the permeability of the core.

Mostly the initial permeability's are below 100 only. Thus these inductors posses with high temperature co-efficient stability. These are mainly applicable in switching power supplies.

Iron Powder Cores

Laminated Core Inductor

These core materials are formed by arranging many numbers of laminations on top of each other. These laminations may be made up of different materials and with different thicknesses. So this construction has more flexibility. These laminations are made up of steel with insulating material between them.

These are arranged parallel to the field to avoid eddy current losses between the laminations. These are used in low frequency detectors. They have high power levels so , they are mostly used at power filtering devices for excitation frequencies above several KHz.

Laminated-core-inductor

Bobbin based Inductor

These are wounded on cylindrical bobbin so these are named as bobbin based inductors. These are mainly used for mounting on printed circuit boards.

It consists of two types of leads they are axial lead and radial lead. Axial lead means lead exits from both sides of the core for horizontal mounting on PC board. Radial lead means lead exits from both sides of the core for vertical mounting on PC board. These are shown below:

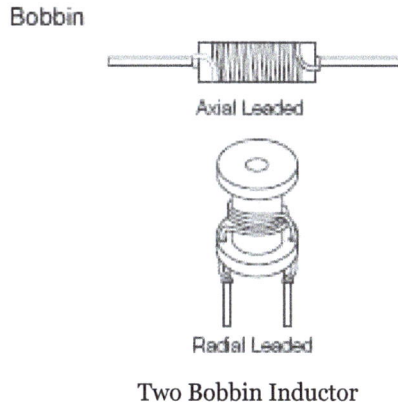

Bobbin

Axial Leaded

Radial Leaded

Two Bobbin Inductor

Toroidal Inductor

Wire wounded on core which has ring or donut shaped surface. These are generally made up of different materials like ferrite, powdered iron and tape wound etc. This inductor has high coupling results between winding and early saturation.

Its arrangement gives minimum loss in magnetic flux which helps to avoid coupling magnetic flux with other devices. It has high energy transferring efficiency and high inductance values at low frequency applications. These inductors mainly used in medical devices, switching regulators, air conditioners, refrigerators, telecommunications and musical instruments etc.

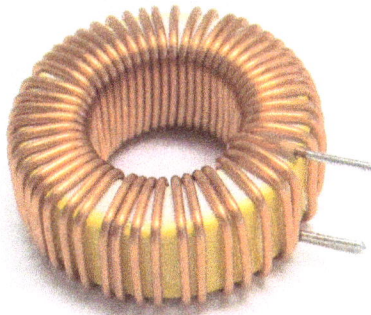

Toroidal Inductor

Multi-layer Ceramic Inductors

The name itself indicates that it consist of multi layers. Simply by adding additional layers of coiled wire that is wound around the central core to the inductor gives multi-layer inductor. Generally for more number of turns in a wire, the inductance is also more.

In these multi-layer inductors not only the inductance of the inductor increases but also the capacitance between the wires also increases. The most advantage of these inductors is by giving the lower operating frequencies also we can get higher inductance results.

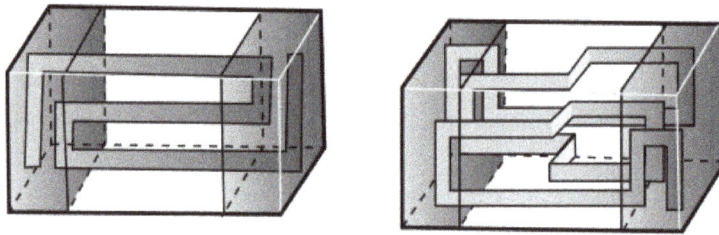

Single layer   Multilayer

Multi-layer Ceramic Inductors

These are having applications at high frequencies to suppress noise, in signal processing modules like wireless LANs, Bluetooth etc. These are also used at mobile communication systems.

Film Inductor

These use a film of conductor on base material. Thus according to the requirement this film is shaped for conductor application. Film inductors in thin size are suitable for DC to DC converters that serve as power supplies in smart phones and mobile devices. The Rf thin film inductor is shown below:

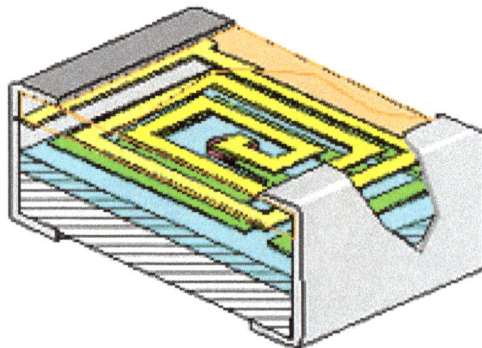

Film Inductor

Variable Inductor

It is formed by moving the magnetic core in and outside of the inductor windings. By this magnetic core we can adjust the inductance value. When we consider a ferrite core inductor , by moving its core inside and outside on which the coil is wounded, variable ferrite core inductor can be formed.

These type of inductors are used in radio and high frequency applications where the tuning is required. These inductors are typically ranged from 10 µH to 100 µH and in present days these are ranged from 10nH to 100 mH.

Variable inductor

Coupled Inductors

The two conductors connected by electromagnetic induction are generally referred as coupled inductors. We already seen that whenever the AC current is flowing in one inductor produces voltage in second inductor gives us mutual inductance phenomenon.

Coupled inductors will work on this phenomenon only. These can isolate two circuits electrically by transferring impedance through the circuit. A transformer is one of the type of coupled inductor.

Coupled Inductors

Molded Inductors

These inductors or molded by plastic or ceramic insulators. These are typically available in bar and cylindrical shapes with wide option of windings.

Molded inductors

Switches

A switch is a component which controls the open-ness or closed-ness of an electric circuit. They allow control over current flow in a circuit (without having to actually get in there and manually cut or splice the wires). Switches are critical components in any circuit which requires user interaction or control.

A switch can only exist in one of two states: open or closed. In the off state, a switch looks like an open gap in the circuit. This, in effect, looks like an open circuit, preventing current from flowing.

In the on state, a switch acts just like a piece of perfectly-conducting wire. This closes the circuit, turning the system "on" and allowing current to flow unimpeded through the rest of the system.

Any switch designed to be operated by a person is generally called a hand switch, and they are manufactured in several varieties:

Toggle Switches

Toggle switch

Toggle switches are actuated by a lever angled in one of two or more positions. The common light switch used in household wiring is an example of a toggle switch. Most toggle switches will come to rest in any of their lever positions, while others have an internal spring mechanism returning the lever to a certain normal position, allowing for what is called "momentary" operation.

Pushbutton Switches

Pushbutton switch

Pushbutton switches are two-position devices actuated with a button that is pressed and released. Most pushbutton switches have an internal spring mechanism returning the button to its "out," or "unpressed," position, for momentary operation. Some pushbutton switches will latch alternately on or off with every push of the button. Other pushbutton switches will stay in their "in," or "pressed," position until the button is pulled back out. This last type of pushbutton switches usually have a mushroom-shaped button for easy push-pull action.

Selector Switches

Selector switch

Selector switches are actuated with a rotary knob or lever of some sort to select one of two or more positions. Like the toggle switch, selector switches can either rest in any of their positions or contain spring-return mechanisms for momentary operation.

Joystick Switch

Joystick switch

A joystick switch is actuated by a lever free to move in more than one axis of motion. One or more of several switch contact mechanisms are actuated depending on which way the lever is pushed, and sometimes by how far it is pushed. The circle-and-dot notation on the switch symbol represents the direction of joystick lever motion required to actuate the contact. Joystick hand switches are commonly used for crane and robot control.

Some switches are specifically designed to be operated by the motion of a machine rather than by the hand of a human operator. These motion-operated switches are commonly called limit switches, because they are often used to limit the motion of a machine by turning off the actuating power to a component if it moves too far. As with hand switches, limit switches come in several varieties:

Limit Switches

Lever actuator limit switch

These limit switches closely resemble rugged toggle or selector hand switches fitted with a lever pushed by the machine part. Often, the levers are tipped with a small roller bearing, preventing the lever from being worn off by repeated contact with the machine part.

Proximity Switches

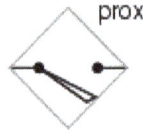

Proximity switch

prox

Proximity switches sense the approach of a metallic machine part either by a magnetic or high-frequency electromagnetic field. Simple proximity switches use a permanent magnet to actuate a sealed switch mechanism whenever the machine part gets close (typically 1 inch or less). More complex proximity switches work like a metal detector, energizing a coil of wire with a high-frequency current, and electronically monitoring the magnitude of that current. If a metallic part (not necessarily magnetic) gets close enough to the coil, the current will increase, and trip the monitoring circuit. The symbol shown here for the proximity switch is of the electronic variety, as indicated by the diamond-shaped box surrounding the switch. A non-electronic proximity switch would use the same symbol as the lever-actuated limit switch.

Another form of proximity switch is the optical switch, comprised of a light source and photocell. Machine position is detected by either the interruption or reflection of a light beam. Optical switches are also useful in safety applications, where beams of light can be used to detect personnel entry into a dangerous area.

The Different Types of Process Switches

In many industrial processes, it is necessary to monitor various physical quantities with switches. Such switches can be used to sound alarms, indicating that a process variable has exceeded normal parameters, or they can be used to shut down processes or equipment if those variables have reached dangerous or destructive levels. There are many different types of process switches:

Speed switch

These switches sense the rotary speed of a shaft either by a centrifugal weight mechanism mounted on the shaft, or by some kind of non-contact detection of shaft motion such as optical or magnetic.

Pressure switch

Gas or liquid pressure can be used to actuate a switch mechanism if that pressure is applied to a piston, diaphragm, or bellows, which converts pressure to mechanical force.

Temperature switch

An inexpensive temperature-sensing mechanism is the "bimetallic strip:" a thin strip of two metals, joined back-to-back, each metal having a different rate of thermal expansion. When the strip heats or cools, differing rates of thermal expansion between the two metals causes it to bend. The bending of the strip can then be used to actuate a switch contact mechanism. Other temperature switches use a brass bulb filled with either a liquid or gas, with a tiny tube connecting the bulb to a pressure-sensing switch. As the bulb is heated, the gas or liquid expands, generating a pressure increase which then actuates the switch mechanism.

Liquid level switch

A floating object can be used to actuate a switch mechanism when the liquid level in an tank rises past a certain point. If the liquid is electrically conductive, the liquid itself can be used as a conductor to bridge between two metal probes inserted into the tank at the required depth. The conductivity technique is usually implemented with a special design of relay triggered by a small amount of current through the conductive liquid. In most cases it is impractical and dangerous to switch the full load current of the circuit through a liquid.

Level switches can also be designed to detect the level of solid materials such as wood chips, grain, coal, or animal feed in a storage silo, bin, or hopper. A common design for this application is a small paddle wheel, inserted into the bin at the desired height, which is slowly turned by a small electric motor. When the solid material fills the bin to that height, the material prevents the paddle wheel from turning. The torque response of the small motor than trips the switch mechanism. Another design uses a "tuning fork" shaped metal prong, inserted into the bin from the outside at the desired height. The fork is vibrated at its resonant frequency by an electronic circuit and magnet/electromagnet coil assembly. When the bin fills to that height, the solid material dampens the vibration of the fork, the change in vibration amplitude and/or frequency detected by the electronic circuit.

Liquid flow switch

Inserted into a pipe, a flow switch will detect any gas or liquid flow rate in excess of a certain threshold, usually with a small paddle or vane which is pushed by the flow. Other flow switches are constructed as differential pressure switches, measuring the pressure drop across a restriction built into the pipe.

Nuclear level switch
(for solid or liquid material)

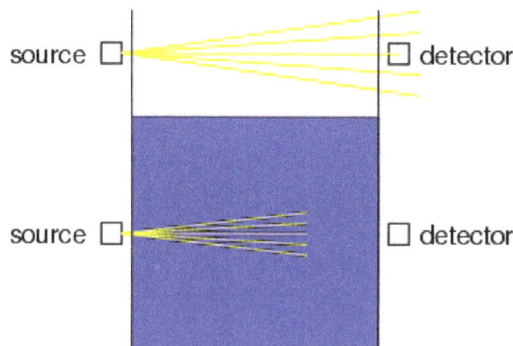

Another type of level switch, suitable for liquid or solid material detection, is the nuclear switch. Composed of a radioactive source material and a radiation detector, the two are mounted across the diameter of a storage vessel for either solid or liquid material. Any height of material beyond the level of the source/detector arrangement will attenuate the strength of radiation reaching the detector. This decrease in radiation at the detector can be used to trigger a relay mechanism to provide a switch contact for measurement, alarm point, or even control of the vessel level.

Both source and detector are outside of the vessel, with no intrusion at all except the radiation flux itself. The radioactive sources used are fairly weak and pose no immediate health threat to operations or maintenance personnel.

As usual, there is usually more than one way to implement a switch to monitor a physical process or serve as an operator control. There is usually no single "perfect" switch for any application, although some obviously exhibit certain advantages over others. Switches must be intelligently matched to the task for efficient and reliable operation.

Batteries

A battery is a collection of one or more cells that go under chemical reactions to create the flow of electrons within a circuit. There is lot of research and advancement going on in battery technology, and as a result, breakthrough technologies are being experienced and used around the world currently. Batteries came into play due to the need to store generated electrical energy. As much as a good amount of energy was being generated, it was important to store the energy so it can be used when generation is down or when there is a need to power standalone devices which cannot be kept tethered to the supply from the mains. Here it should be noted that only DC can be stored in the batteries, AC current can't be stored.

Battery cells are usually made up of three main components:

1. The Anode (Negative Electrode)

2. The Cathode (Positive Electrode)

3. The electrolytes

The anode is a negative electrode that produces electrons to the external circuit to which the battery is connected. When batteries are connected, an electron build up is initiated at the anode which causes a potential difference between the two electrodes. The electrons naturally then try to redistribute themselves, this is prevented by the electrolyte, so when an electrical circuit is connected, it provides a clear path for the electrons to move from the anode to the cathode thereby powering the circuit to which it is connected.

Types of Batteries

Batteries generally can be classified into different categories and types, ranging from chemical composition, size, form factor and use cases, but under all of these are two major battery types:

1. Primary Batteries

2. Secondary Batteries

1. Primary Batteries

Primary batteries are batteries that cannot be recharged once depleted. Primary batteries are made of electrochemical cells whose electrochemical reaction cannot be reversed.

Primary batteries exist in different forms ranging from coin cells to AA batteries. They are commonly used in standalone applications where charging is impractical or impossible. A good example of which is in military grade devices and battery powered equipment. It will be impractical to use rechargeable batteries as recharging a battery will be the last thing in the mind of the soldiers. Primary batteries always have high specific energy and the systems in which they are used are always designed to consume low amount of power to enable the battery last as long as possible.

Primary Batteries

Some other examples of devices using primary batteries include: Pace makers, Animal trackers, Wrist watches, remote controls and children toys to mention a few.

The most popular type of primary batteries are alkaline batteries. They have a high specific energy and are environmentally friendly, cost-effective and do not leak even when fully discharged. They can be stored for several years, have a good safety record and can be carried on an aircraft without being subject to UN Transport and other regulations. The only downside to alkaline batteries is the low load current, which limits its use to devices with low current requirements like remote controls, flashlights and portable entertainment devices.

2. Secondary Batteries

Secondary batteries are batteries with electrochemical cells whose chemical reactions can be reversed by applying a certain voltage to the battery in the reversed direction. Also referred to as rechargeable batteries, secondary cells unlike primary cells can be recharged after the energy on the battery has been used up.

They are typically used in high drain applications and other scenarios where it will be either too expensive or impracticable to use single charge batteries. Small capacity secondary batteries are used to power portable electronic devices like mobile phones, and other gadgets and appliances while heavy-duty batteries are used in powering diverse electric vehicles and other high drain applications like load levelling in electricity generation. They are also used as standalone power sources alongside Inverters to supply electricity. Although the initial cost of acquiring rechargeable batteries is always a whole lot higher than that of primary batteries but they are the most cost-effective over the long-term.

Secondary batteries can be further classified into several other types based on their chemistry. This is very important because the chemistry determines some of the attributes of the battery including its specific energy, cycle life, shelf life, and price to mention a few.

There are basically four major chemistries for rechargeable batteries:

Nickel-Cadmium Batteries

The nickel–cadmium battery (NiCd battery or NiCad battery) is a type of rechargeable battery which is developed using nickel oxide hydroxide and metallic cadmium as electrodes. Ni-Cd batteries excel at maintaining voltage and holding charge when not in use. However, NI-Cd batteries easily fall a victim of the dreaded "memory" effect when a partially charged battery is recharged, lowering the future capacity of the battery.

Nickel – Cadmium battery

In comparison with other types of rechargeable cells, Ni-Cd batteries offer good life cycle and performance at low temperatures with a fair capacity but their most significant advantage will be their ability to deliver their full rated capacity at high discharge rates. They are available in different sizes including the sizes used for alkaline batteries, AAA to D. Ni-Cd cells are used individual or assembled in packs of two or more cells. The small packs are used in portable devices, electronics and toys while the bigger ones find application in aircraft starting batteries, Electric vehicles and standby power supply.

Some of the properties of Nickel-Cadmium batteries are listed below.

Specific Energy: 40-60W-h/kg

Energy Density: 50-150 W-h/L

Specific Power: 150W/kg

Charge/discharge efficiency: 70-90%

Self-discharge rate: 10%/month

Cycle durability/life: 2000cycles

Nickel-Metal Hydride Batteries

Nickel-metal hydride (Ni-MH) is another type of chemical configuration used for rechargeable batteries. The chemical reaction at the positive electrode of batteries is similar to that of the nickel–cadmium cell (NiCd), with both battery type using the same nickel oxide hydroxide (NiOOH). However, the negative electrodes in Nickel-Metal Hydride use a hydrogen-absorbing alloy instead of cadmium which is used in NiCd batteries.

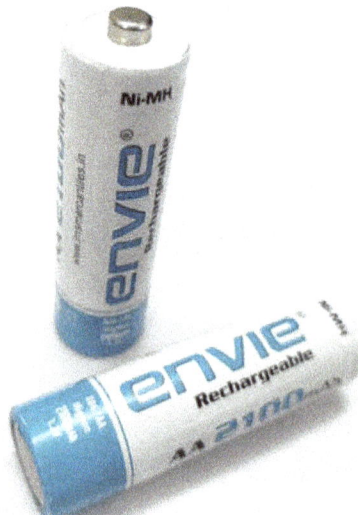

Ni-MH Battery

NiMH batteries find application in high drain devices because of their high capacity and energy density. A NiMH battery can possess two to three times the capacity of a NiCd battery of the same size, and its energy density can approach that of a lithium-ion battery. Unlike the NiCd chemistry, batteries based on the NiMH chemistry are not susceptible to the "memory" effect that NiCads experience.

Below are some of the properties of batteries based on the Nickel-metal hydride chemistry:

Specific Energy: 60-120h/kg

Energy Density: 140-300 Wh/L

Specific Power: 250-1000 W/kg

Charge/discharge efficiency: 66% - 92%

Self-discharge rate: 1.3-2.9%/month at 200C

Cycle Durability/life: 180 -2000

Lithium-ion Batteries

Lithium ion batteries are one of the most popular types of rechargeable batteries. They are found in different portable appliances including mobile phones, smart devices and several other battery appliances used at home. They also find applications in aerospace and military applications due to their lightweight nature.

Lithium-Ion Battery

Lithium-ion batteries are a type of rechargeable battery in which lithium ions from the negative electrode migrate to the positive electrode during discharge and migrate back to the negative electrode when the battery is being charged. Li-ion batteries use an intercalated lithium compound as one electrode material, compared to the metallic lithium used in non-rechargeable lithium batteries.

Lithium-ion batteries generally possess high energy density, little or no memory effect and low self-discharge compared to other battery types. Their chemistry alongside performance and cost vary across different use cases for example, Li-ion batteries used in handheld electronic devices are usually based on lithium cobalt oxide ($LiCoO_2$) which provides high energy density and low safety risks when damaged while Li-ion batteries based on Lithium iron phosphate which offer a lower energy density are safer due to a reduced likelihood of unfortunate events happening are widely used in powering electric tools and medical equipment. Lithium ion batteries offer the best performance to weight ratio with the lithium sulphur battery offering the highest ratio.

Some of the attributes of lithium ion batteries are listed below:

Specific Energy: 100: 265W-h/kg

Energy Density: 250: 693 W-h/L

Specific Power: 250: 340 W/kg

Charge/discharge percentage: 80-90%

Cycle Durability: 400: 1200 cycles

Nominal cell voltage: NMC 3.6/3.85V

Lead-acid Batteries

Lead-acid batteries are a low-cost reliable power workhorse used in heavy duty applications. They are usually very large and because of their weight, they're always used in non-portable applications such as solar-panel energy storage, vehicle ignition and lights, backup power and load levelling in power generation/distribution. The lead-acid is the oldest type of rechargeable battery and still very relevant and important into today's world. Lead-acid batteries have very low energy to volume and energy to weight ratios but it has a relatively large power to weight ratio and as a result can supply huge surge currents when needed. These attributes alongside its low cost makes these batteries attractive for use in several high current applications like powering automobile starter motors and for storage in backup power supplies.

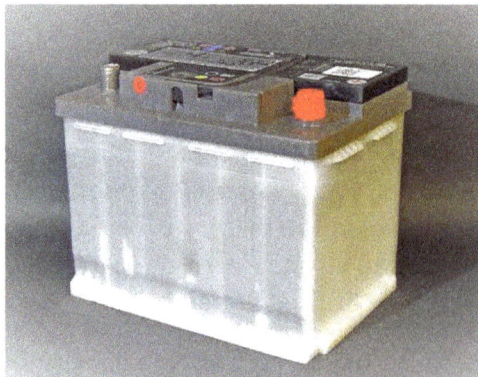

Lead-acid Batteries

Each of these batteries has its area of best fit and the image below is to help choose between them.

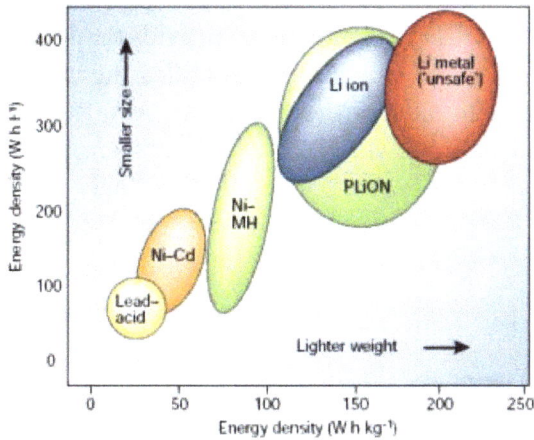

Selecting the right battery for your application

One of the main problems hindering technology revolutions like IoT is power, battery life affects the successful deployment of devices that require long battery life and even though several power management techniques are being adopted to make the battery last longer, a compatible battery must still be selected to achieve the desired outcome.

Below are some factors to consider when selecting the right type of battery for your project:

1. Energy Density: The energy density is the total amount of energy that can be stored per unit mass or volume. This determines how long your device stays on before it needs a recharge.

2. Power Density: Maximum rate of energy discharge per unit mass or volume. Low power: laptop, i-pod. High power: power tools.

3. Safety: It is important to consider the temperature at which the device you are building will work. At high temperatures, certain battery components will breakdown and can undergo exothermic reactions. High temperatures generally reduces the performance of most batteries.

4. Life cycle durability: The stability of energy density and power density of a battery with repeated cycling (charging and discharging) is needed for the long battery life required by most applications.

5. Cost: Cost is an important part of any engineering decisions you will be making. It is important that the cost of your battery choice is commensurate with its performance and will not increase the overall cost of the project abnormally.

Voltage Source

A voltage source, such as a battery or generator, provides a potential difference (voltage) between two points within an electrical circuit allowing current to flowing around it. Remember that voltage can exist without current. A battery is the most common voltage source for a circuit with the voltage that appears across the positive and negative terminals of the source being called the terminal voltage.

Ideal Voltage Source

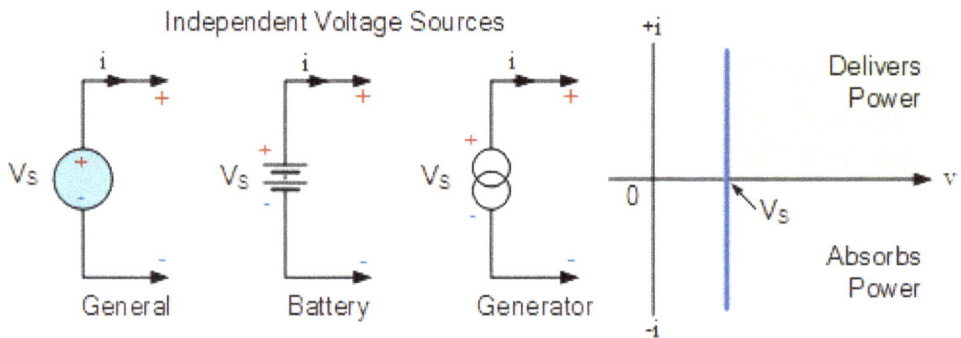

An ideal voltage source is defined as a two terminal active element that is capable of supplying and maintaining the same voltage, (v) across its terminals regardless of the current, (i) flowing through it. In other words, an ideal voltage source will supply a constant voltage at all times regardless of the value of the current being supplied producing an I-V characteristic represented by a straight line.

Then an ideal voltage source is known as an Independent Voltage Source as its voltage does not depend on either the value of the current flowing through the source or its direction but is determined solely by the value of the source alone. So for example, an automobile battery has a 12V terminal voltage that remains constant as long as the current through it does not become to high, delivering power to the car in one direction and absorbing power in the other direction as it charges.

On the other hand, a Dependent Voltage Source or controlled voltage source, provides a voltage supply whose magnitude depends on either the voltage across or current flowing through some other circuit element. A dependent voltage source is indicated with a diamond shape and are used as equivalent electrical sources for many electronic devices, such as transistors and operational amplifiers.

Connecting Voltage Sources Together

Ideal voltage sources can be connected together in both parallel or series the same as for any circuit element. Series voltages add together while parallel voltages have the same value. Note that unequal ideal voltage sources cannot be connected directly together in parallel.

Voltage Source in Parallel

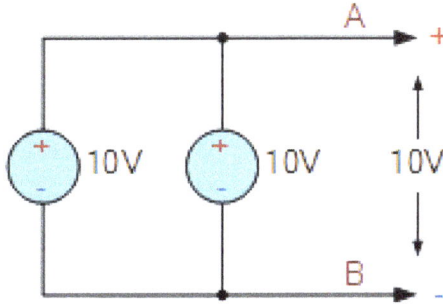

While not best practice for circuit analysis, ideal voltage sources can be connected in parallel provided they are of the same voltage value. Here in this example, two 10 volt voltage sources are combined to produce 10 volts between terminals A and B. Ideally, there would be just one single voltage source of 10 volts given between terminals A and B.

What is not allowed or is not best practice, is connecting together ideal voltage sources that have different voltage values as shown, or are short-circuited by an external closed loop or branch.

Badly Connected Voltage Sources

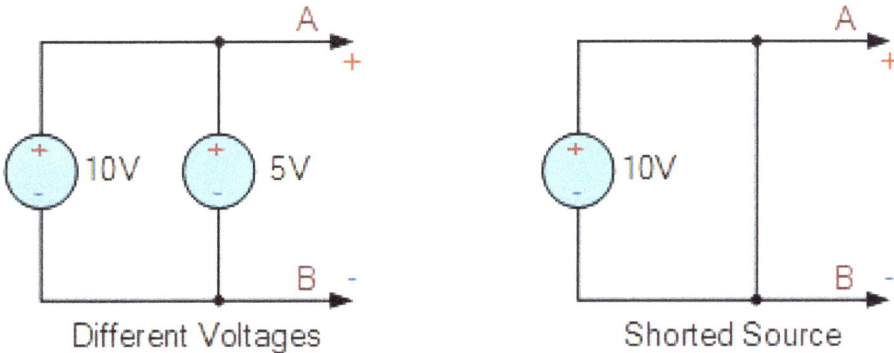

Different Voltages

Shorted Source

However, when dealing with circuit analysis, voltage sources of different values can be used providing there are other circuit elements in between them to comply with Kirchoff's Voltage Law, KVL.

Unlike parallel connected voltage sources, ideal voltage sources of different values can be connected together in series to form a single voltage source whose output will be the algebraic addition or subtraction of the voltages used. Their connection can be as: series-aiding or series-opposing voltages as shown.

Voltage Source in Series

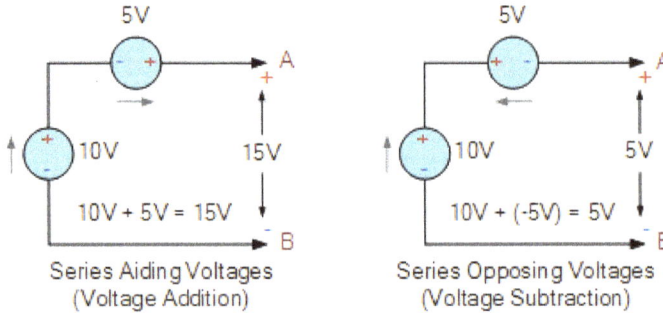

5V   5V

A   A

10V   15V   10V   5V

10V + 5V = 15V   10V + (-5V) = 5V

B   B

Series Aiding Voltages   Series Opposing Voltages
(Voltage Addition)   (Voltage Subtraction)

Series aiding voltage sources are series connected sources with their polarities connected so that the plus terminal of one is connected to the negative terminal of the next allowing current to flow in the same direction. In the example above, the two voltages of 10V and 5V of the first circuit can be added, for a V_S of 10 + 5 = 15V. So the voltage across terminals A and B is 15 volts.

Series opposing voltage sources are series connected sources which have their polarities connected so that the plus terminal or the negative terminals are connected together as shown in the second circuit above. The net result is that the voltages are subtracted from each other. Then the two voltages of 10V and 5V of the second circuit are subtracted with the smaller voltage subtracted from the larger voltage. Resulting in a V_S of 10 - 5 = 5V.

The polarity across terminals A and B is determined by the larger polarity of the voltage sources, in this example terminal A is positive and terminal B is negative resulting in +5 volts. If the series-opposing voltages are equal, the net voltage across A and B will be zero as one voltage balances out the other. Also any currents (I) will also be zero, as without any voltage source, current cannot flow.

Voltage Source Example

Two series aiding ideal voltage sources of 6 volts and 9 volts respectively are connected together to supply a load resistance of 100 Ohms. Calculate: the source voltage, V_S, the load current through the resistor, I_R and the total power, P dissipated by the resistor. Draw the circuit.

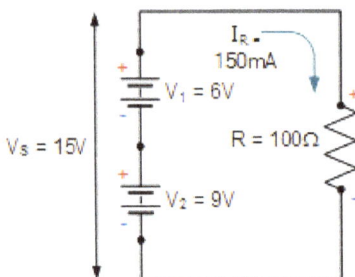

I_R =
150mA
V_1 = 6V

V_S = 15V   R = 100Ω

V_2 = 9V

$V_S = V_1 + V_2 = 6 + 9 = 15V$

$I_R = \dfrac{V_S}{R} = \dfrac{15V}{100\Omega} = 150mA$

$P_R = I^2R = 0.15^2 \times 100 = 2.25W$

$$V_s = V_1 + V_2 = 6 + 9 = 15\,V$$

$$I_R = \frac{V_S}{R} = \frac{15\,V}{100\,\Omega} - 150\,mA$$

$$P_R = I^2R = 0.15^2 \times 100 = 2.25\,W$$

Thus, V_s = 15V, I_R = 150mA or 0.15A, and P_R = 2.25W.

Practical Voltage Source

We have seen that an ideal voltage source can provide a voltage supply that is independent of the current flowing through it, that is, it maintains the same voltage value always. This idea may work well for circuit analysis techniques, but in the real world voltage sources behave a little differently as for a practical voltage source, its terminal voltage will actually decrease with an increase in load current.

As the terminal voltage of an ideal voltage source does not vary with increases in the load current, this implies that an ideal voltage source has zero internal resistance, RS = 0. In other words, it is a resistorless voltage source. In reality all voltage sources have a very small internal resistance which reduces their terminal voltage as they supply higher load currents.

For non-ideal or practical voltage sources such as batteries, their internal resistance (R_s) produces the same effect as a resistance connected in series with an ideal voltage source as these two series connected elements carry the same current as shown.

Ideal and Practical Voltage Source

You may have noticed that a practical voltage source closely resembles that of a Thevenin's equivalent circuit as Thevenin's theorem states that "any linear network containing resistances and sources of emf and current may be replaced by a single voltage source, V_s in series with a single resistance, R_s". Note that if the series source resistance is low, the voltage source is ideal. When the source resistance is infinite, the voltage source is open-circuited.

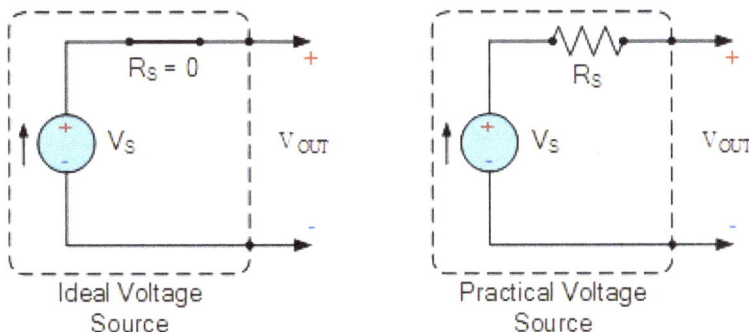

Ideal Voltage
Source

Practical Voltage
Source

In the case of all real or practical voltage sources, this internal resistance, RS no matter how small has an effect on the I-V characteristic of the source as the terminal voltage falls off with an increase in load current. This is because the same load current flows through R_s.

Ohms law tells us that when a current (i) flows through a resistance, a voltage drop is produce across the same resistance. The value of this voltage drop is given as $i*R_s$. Then VOUT will equal the ideal voltage source, V_s minus the $i*R_s$ voltage drop across the resistor. Remember that in the case of an ideal source voltage, RS is equal to zero as there is no internal resistance, therefore the terminal voltage is same as V_s.

Then the voltage sum around the loop given by Kirchhoff's voltage law, KVL is: $V_{OUT} = V_s - i*R_s$. This equation can be plotted to give the I-V characteristics of the actual output voltage. It will give a straight line with a slope $-R_s$ which intersects the vertical voltage axis at the same point as V_s when the current i = 0 as shown.

Practical Voltage Source Characteristics

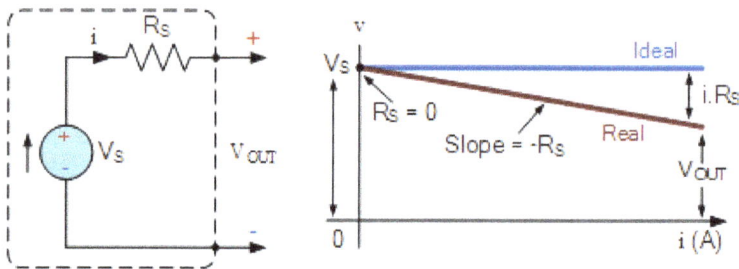

Therefore, all ideal voltage sources will have a straight line I-V characteristic but non-ideal or real practical voltage sources will not but instead will have an I-V characteristic that is slightly angled down by an amount equal to $i*R_s$ where R_s is the internal source resistance (or impedance). The I-V characteristics of a real battery provides a very close approximation of an ideal voltage source since the source resistance R_s is usually quite small.

The decrease in the angle of the slope of the I-V characteristics as the current increases is known as regulation. Voltage regulation is an important measure of the quality of a practical voltage source as it measures the variation in terminal voltage between no load, that is when $I_L = 0$, (an open-circuit) and full load, that is when I_L is at maximum, (a short-circuit).

Voltage Source Example

A battery supply consists of an ideal voltage source in series with an internal resistor. The voltage and current measured at the terminals of the battery were found to be V_{OUT1} = 130V at 10A, and V_{OUT2} = 100V at 25A. Calculate the voltage rating of the ideal voltage source and the value of its internal resistance. Draw the I-V characteristics.

Firstly lets define in simple "simultaneous equation form", the two voltage and current outputs of the battery supply given as: V_{OUT1} and V_{OUT2}.

$$V_{OUT} = V_S - iR_S$$
$$V_{OUT1} = 130 = V_S + 10R_S$$
$$V_{OUT2} = 100 = V_S + 25R_S$$

As with have the voltages and currents in a simultaneous equation form, to find V_S we will first multiply V_{OUT1} by five, and V_{OUT2} by two, as shown to make the value of the two currents (i) the same for both equations,

$$V_{OUT1} = 130 = V_S + 10R_S \dots\dots \times 5$$
$$V_{OUT2} = 100 = V_S + 25R_S \dots\dots \times 2$$

$$V_{OUT1} = 650 = 5V_S + 50R_S$$
$$V_{OUT2} = 200 = 2V_S + 50R_S$$

Having made the coefficients for R_S the same by multiplying through with the previous constants, we now multiply the second equation V_{OUT2} by minus one, (-1) to allow for the subtraction of the two equations so that we can solve for VS as shown.

$$V_{OUT1} = 650 = 5V_S + 50R_S$$
$$V_{OUT2} = 200 = 2V_S + 50R_S \dots\dots \times -1$$

$$V_{OUT1} = 650 = 5V_S + 50R_S$$
$$V_{OUT2} = -200 = -2V_S - 50R_S$$

Rearrange to give

$$650 - 200 = \left(5V_S - 2V_S\right) + \left(50R_S - 50R_S\right)$$
$$450 = 3V_S + 0$$
$$\therefore V_S = \frac{450}{3} = 150V$$

Knowing that the ideal voltage source, V_S is equal to 150 volts, we can use this value for equation V_{OUT1} (or V_{OUT2} if so wished) and solve to find the series resistance, R_S.

$$V_{OUT1} = 130 = V_S + 10R_S$$
$$V_S = 150V$$
$$130 = 150 + 10R_S$$
$$\therefore R_S = \frac{150 - 130}{10} = 2\Omega$$

Then for our simple example, the batteries internal voltage source is calculated as: $V_S = 150$ volts, and its internal resistance as: $R_S = 2\Omega$. The I-V characteristics of the battery are given as:

Battery I-V Characteristics

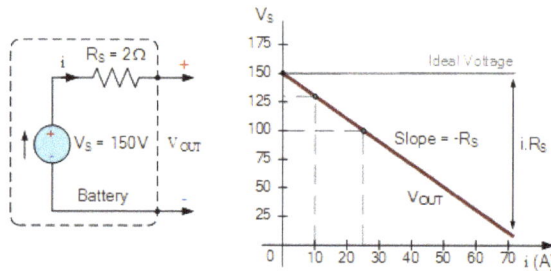

Dependent Voltage Source

Unlike an ideal voltage source which produces a constant voltage across its terminals regardless of what is connected to it, a controlled or dependent voltage source changes its terminal voltage depending upon the voltage across, or the current through, some other element connected to the circuit, and as such it is sometimes difficult to specify the value of a dependent voltage source, unless you know the actual value of the voltage or current on which it depends.

Dependent voltage sources behave similar to the electrical sources we have looked at so far, both practical and ideal (independent) the difference this time is that a dependent voltage source can be controlled by an input current or voltage. A voltage source that depends on a voltage input is generally referred to as a Voltage Controlled Voltage Source or VCVS. A voltage source that depends on a current input is referred to as a Current Controlled Voltage Source or CCVS.

Ideal dependent sources are commonly used in the analysing the input/output characteristics or the gain of circuit elements such as operational amplifiers, transistors and integrated circuits. Generally, an ideal voltage dependent source, either voltage or current controlled is designated by a diamond-shaped symbol as shown.

Dependent Voltage Source Symbols

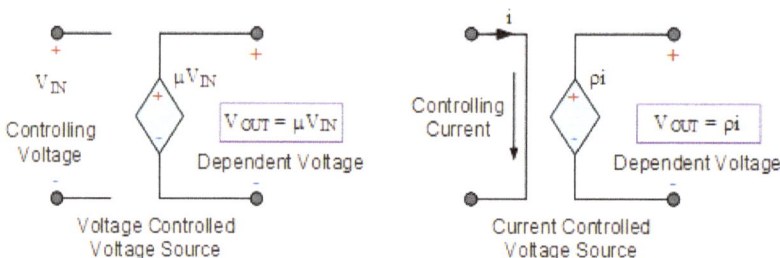

An ideal dependent voltage-controlled voltage source, VCVS, maintains an output voltage equal to some multiplying constant (basically an amplification factor) times the controlling voltage present elsewhere in the circuit. As the multiplying constant is, well, a constant, the controlling voltage, VIN will determine the magnitude of the output voltage, V_{OUT}. In other words, the output voltage "depends" on the value of input voltage making it a dependent voltage source and in many ways, an ideal transformer can be thought of as a VCVS device with the amplification factor being its turns ratio.

Then the VCVS output voltage is determined by the following equation: $V_{OUT} = \mu V_{IN}$. Note that the multiplying constant μ is dimensionless as it is purely a scaling factor because $\mu = V_{OUT}/V_{IN}$, so its units will be volts/volts.

An ideal dependent current-controlled voltage source, CCVS, maintains an output voltage equal to some multiplying constant (rho) times a controlling current input generated elsewhere within the connected circuit. Then the output voltage "depends" on the value of the input current, again making it a dependent voltage source.

As a controlling current, IIN determines the magnitude of the output voltage, V_{OUT} times the magnification constant ρ (rho), this allows us to model a current-controlled voltage source as a trans-resistance amplifier as the multiplying constant, ρ gives us the following equation: $V_{OUT} = \rho I_{IN}$. This multiplying constant ρ (rho) has the units of Ohm's because $\rho = V_{OUT}/I_{IN}$, and its units will therefore be volts/amperes.

Voltage Source

We have seen here that a Voltage Source can be either an ideal independent voltage source, or a controlled dependent voltage source. Independent voltage sources supply a constant voltage that does not depend on any other quantity within the circuit. Ideal independent sources can be batteries, DC generators or time-varying AC voltage supplies from alternators.

Independent voltage sources can be modelled as either an ideal voltage source, ($R_S = 0$) where the output is constant for all load currents, or a non-ideal or practical, such as a battery with a resistance connected in series with the circuit to represent the internal resistance of the source. Ideal voltage sources can be connected together in parallel only if they are of the same voltage value. Series-aiding or series-opposing connections will affect the output value.

Also for solving circuit analysis and complex theorems, voltage sources become short-circuited sources making their voltage equal to zero to help solve the network. Note also that voltage sources are capable of both delivering or absorbing power.

Ideal dependent voltage sources represented by a diamond-shaped symbol, are dependent on, and are proportional too an external controlling voltage or current. The multiplying constant, μ for a V_{cvs} has no units, while the multiplying constant ρ for a C_{cvs} has units of Ohm's. A dependent voltage source is of great interest to model electronic devices or active devices such as operational amplifiers and transistors that have gain.

Wires

Electrical wire is the medium through which electricity is carried to and through each individual home that uses electrical power. It is made of a metal that easily conducts electricity, usually copper, in a plastic sheath called an insulator. There are various different types of this wire, each suited to certain loads and conditions.

Some factors that will affect your choice of electrical wiring include color, label information and applications. The information printed on the wire covering is all that you need to choose the correct wire for your home. Here's some detailed information on the various features of electrical wire, which will help you choose the correct composition.

Types of Wires

There are mainly 5 types of wire:

- Triplex Wires: Triplex wires are usually used in single-phase service drop conductors, between the power pole and weather heads. They are composed of two insulated aluminum wires wrapped with a third bare wire which is used as a common neutral. The neutral is usually of a smaller gauge and grounded at both the electric meter and the transformer.

- Main Feeder Wires: Main power feeder wires are the wires that connect the service weather head to the house. They're made with stranded or solid THHN wire and the cable installed is 25% more than the load required.

- Panel Feed Wires: Panel feed cables are generally black insulated THHN wire. These are used to power the main junction box and the circuit breaker panels. Just like main power feeder wires, the cables should be rated for 25% more than the actual load.

- Non-metallic Sheathed Wires: Non-metallic sheath wire, or Romex, is used in most homes and has 2-3 conductors, each with plastic insulation, and a bare ground wire. The individual wires are covered with another layer of non-metallic sheathing. Since it's relatively cheaper and available in ratings for 15, 20 and 20 amps, this type is preferred for in-house wiring.

- Single Strand Wires: Single strand wire also uses THHN wire, though there are other variants. Each wire is separate and multiple wires can be drawn together through a pipe easily. Single strand wires are the most popular choice for layouts that use pipes to contain wires.

Color Codes: Different color wires serve different purposes, like,

- Black: Hot wire, for switches or outlets.

- Red: Hot wire, for switch legs. Also for connecting wire between 2 hardwired smoke detectors.

- Blue and Yellow: Hot wires, pulled in conduit. Blue for 3-4 way switch application, and yellow for switch legs to control fan, lights etc.

- White: Always neutral.

- Green and Bare Copper: Only for grounding.

Size of Wires: Each application requires a certain wire size for installation, and the right size for a specific application is determined by the wire gauge. Sizing of wire is done by the American wire gauge system. Common wire sizes are 10, 12 and 14 – a higher number means a smaller wire size, and affects the amount of power it can carry. For example, a low-voltage lamp cord with 10 Amps will require 18-gauge wire, while service panels or subpanels with 100 Amps will require 2-gauge wire.

Wire Lettering: The letters THHN, THWN, THW and XHHN represent the main insulation types of individual wires. These letters depict the following NEC requirements:

- T – Thermoplastic insulation;

- H – Heat resistance;

- HH – High heat resistance (up to 194° F);

- W – Suitable for wet locations;

- N – Nylon coating, resistant to damage by oil or gas;

- X – Synthetic polymer that is flame-resistant.

Wire Gauge, Ampacity and Wattage Load – To determine the correct wire, it is important to understand what ampacity and wattage a wire can carry per gauge. Wire gauge is the size of the wire, ampacity is how much electricity can flow through the wire and wattage is the load a wire can take, which is always mentioned on the appliances.

Diodes

A diode is an electric device that permits the flow of current only in one direction and restricts the flow in the opposite direction. The most ordinary sort of diode in current circuit design is the semi-conductor diode, even though additional diode technologies are present. The word "diode" is traditionally aloof for tiny signal appliances, I ≤ 1 A. When a diode is positioned in a simple battery lamp circuit, then the diode will either permit or stop flow of current through the lamp, all this depend on the polarization of the volts applied. There are various sorts of diode but their fundamental role is identical. The most ordinary kind of diode is silicon diode; it is placed in a glass cylinder.

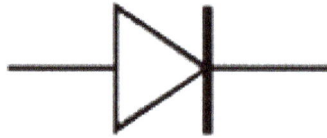

The Symbol of Diode is Represented as Follows:

Diode Operation

A diode starts its operations when a voltage signal applies across its terminals. A DC volt is applied so that diode starts its operation in a circuit and this is known as Biasing. Diode is similar to a switch which is one way, hence it can be either in conduction more or non-conduction mode. "ON" mode of the diode, is attained by forward biasing, which simply means that higher or positive potential is applied on the anode and on the cathode, negative or lower potential is applied of a diode. Whereas the "OFF" mode of the diode is attained with the aid of reverse biasing which simply means that higher or positive potential is applied on the cathode and on the anode, negative or lower potential is applied of a diode.

In the "ON" situation the practical diode provides forward resistance. A diode needs forward bias voltage to get in the "ON" mode this is known as cut-in-voltage. Whereas the diode initiates conducting in reverse biased manner when reverse bias voltage goes beyond its limit and this is known as breakdown voltage. The diode rests in OFF mode when no voltage is applicable across it.

Function of Diode

The key function of a diode is to obstruct the flow of current in one direction and permit the flow of current in the other direction. Current passing through the diode is known as forward current whereas the current blocked by the diode is known as the reverse current.

Diode Equation

The equation of diode expresses the current flow via diode as a function of voltage. The ideal diode equation is:

$$I = I_0 \left(e^{\frac{qv}{kT}} - 1 \right)$$

Where,

- I – Stands for the net current passing through a diode;

- I_0 - Stands for dark saturation current, the diode see page current density in the deficiency of light;

- V – Stands for applied voltage across the terminals of the diode;

- q – Stands for fixed value of electron charge;

- k – Stands for Boltzmann's constant;

- T – Stands for fixed temperature (K).

Diode Circuits

The basic aim behind this study is to show how diodes can be employed in circuits. Now let us analyze a simple diode circuit. Keep in mind what we learnt about ideal diodes. Here we are assuming that diode is ideal:

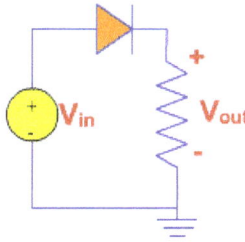

1. When diode is in ON mode, no voltage is there across it; hence it acts like a short circuit.

2. Whereas when diode is in OFF mode, there is zero current, hence it behaves like an open circuit.

3. From the above two conditions, either one can take place at a time. This helps us to check out what will happen in any circuit with diodes.

Diode Characteristics

Diodes have attributes that allow them to carry out a number of electronic functions. Three vital characteristics of diodes are as follows:

- Forward Voltage Drop- forward bias about seven volts.

- Reverse Voltage Drop- Weakened layer broadens, generally the applied voltage.

- Reverse breakdown voltage- reverse voltage drop that'll force flow of current and in maximum cases demolish the diodes.

Application of Diodes

Diodes are employed in a variety of applications such as clipper, rectification, clamper, comparator, voltage multiplier, filters, sampling gates, etc.

Rectification: Rectification symbolizes the alteration of AC volt into DC volt. Some of the common examples of rectification circuits are- FWR (full wave rectifier), bridge rectifier & HWR (half wave rectifier).

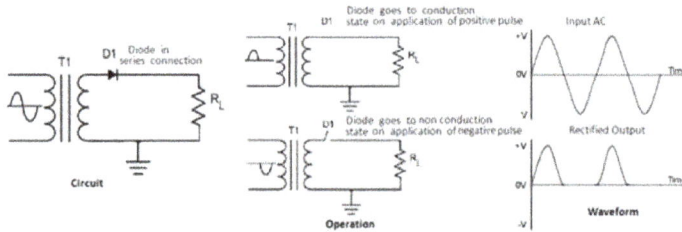

Half wave rectifier

Clipper: Diode can be employed to trim down some fraction of pulse devoid of deforming the left over fraction of the waveform,

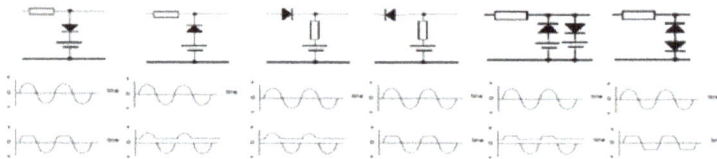

Clipper

Clamper: A clamping circuit limits the level of voltage to go beyond a limit by changing the DC level. The crest to crest is not influenced by clamping. Capacitors, resistors & diodes all are used to create clamping circuits.

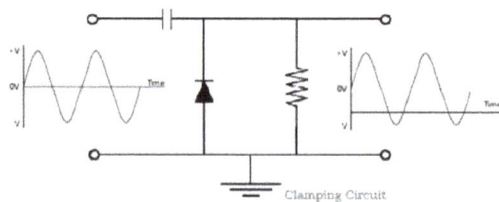

Clamper

Types of Diodes

All sort of diodes are dissimilar in means of construction, characteristics & applications. Following are some of the types of diodes:

- Small Current Diode
- Large Signal or Large Current Diodes

- Zener Diodes

- LED (Light Emitting Diodes)

- Photodiodes

- Constant Current Diodes

- Schottky Diode

- Shockley Diode

- Step Recovery Diodes

- Tunnel Diodes

- Varactor Diodes

- PIN Diodes

- LASER Diode

- Transient Voltage Suppression Diodes

- Gold Doped Diodes

- Super Barrier Diodes

- Point Contact Diodes

- Peltier Diodes

- Gunn Diode

- Vacuum Diodes

- Avalanche Diode

- Crystal Diode

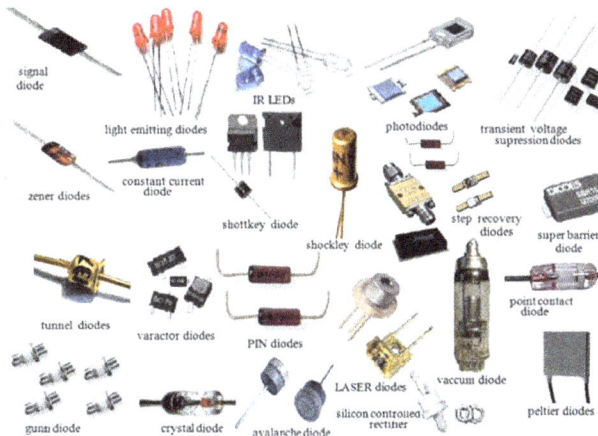

Zener Diode

Zener diode works in reverse bias situation when the voltage attains the breakdown peak. An even voltage can be attained by insertion of a resistor across it to limit the flow of current. This Zener diode is employed to give reference voltage in power supplying circuits.

Symbol of Zener Diode

Zener Diode Characteristics

Special diodes such as zener diodes are intended & manufactured to function in the opposite direction without being broken.

The zener diode acts like a common silicon diode, during the forward bias.

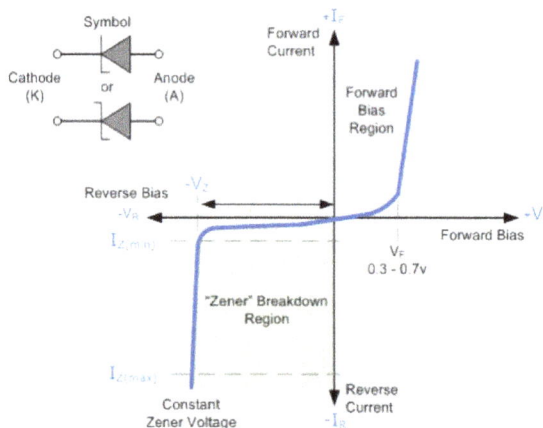

2. Changeable quantity of reverse current can go through the diode devoid of destructing it. The V_z (zener voltage) or breakdown voltage across the diode upholds comparatively steady.

3. Producers rate zener diodes as per their zener voltage value and the highest PD (Power Dissipation) i.e. at 25° C. This provides a signal of the highest I_R (reverse current), that a diode securely carries out.

Zener Diode Applications

Zener diodes have many applications in transistor circuitry. Here we are discussing various vital points in Zener diode applications:

- Zener Diode Shunt Regulator- This diode is commonly employed as a Voltage Regulator or Shunt Regulator.

- Meter Protection- This diode may also come across its functions in meter security.

- Zener Diode as Peak Clipper- This diodes can be employed to cut off the maximum value of incoming waveform.

- Switching operation- This diode can generate an unexpected alteration from low to high current, so it is functional in switching applications. It is relatively speedy in switching processes.

P-N Junction Diode

The simplest semi-conductor appliance is named as p-n junction diode. It is a 2 terminal, one-sided, bipolar repairing device that conducts just in 1 direction. The common diodes are employed in the following areas:

- Modification in power supplying circuits.

- Mining of modulation from broadcasting signals in a radio recipient and in security circuits where big transient currents may emerge on ICs or low current transistors in interfacing with relays or other soaring power appliances.

- Employed in series with power inputs to electric circuits where just 1 positive or negative polarity voltage is required.

P - N Diode Symbol   Internal Structure of P-N Diode

Schottky Diode

Symbol of Schottky Diode

A schottky diode is also identified as hot electron diode or schottky barrier diode. It has a stumpy forward voltage drop with extremely quick switching features. Schottky barrier diode is a solid state diode in which the joint is shaped by the contact of a metal semiconductor. The hot electron diode is generally made-up from N-kind of material or slightly doped substrate and a metal like aluminum. An exceedingly doped substrate may provide a reduced ohmic contact and an elevated series of resistance, therefore not sensible. The metal has been dissolved or stammered on top of the semi-conductor. The hot carrier diode is another kind of schottky diode. Some of the common kinds of schottky diode are- 1n5818, 1n5817, 1n5819.

Benefits of Schottky Diode

- Rapid switching with quicker revival time.

- Lower intersection capacitance.

- Less power loss due to low turn on/forward voltage.

Tunnel Diode

A tunnel diode is a highly conducting two terminal p-n jnction diode doped heavily approximately 1000 times upper than a usual junction diode. A tunnel diode is also named as Esaki diode, it's named after Esaki diode who is a Nobel prize winner in physics for discovering electron tunneling outcome employed in these diodes. Tunnel diodes are helpful in several circuit purposes like in- microwave oscillation, binary memory & microwave amplification. Tunnel diodes are generally made-up from gallium or germanium or gallium arsenide. These all comprise tiny prohibited energy breaks and elevated ion motilities.

Silicon is not utilized in the manufacturing of tunnel diodes owing to its low value (I_p, I_v).

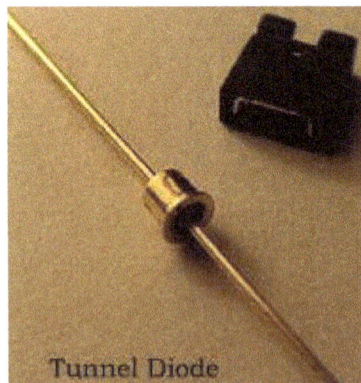

Tunnel Diode

Light Emitting Diode or LED

LED is a semiconductor appliance that produces visible light beams or infrared light beams when an electric current is passed through it. Visible LEDs can be seen in several electronic devices such as microwaves' number display light, brake lights, and even cameras to make use of Infrared LEDs. In LEDs (light emitting diodes) light is created by a solid situation procedure which is named as electroluminescence.

LED   LED Circuit Symbol

Light emitting diodes are available in various colors like- orange, red, yellow, amber, green, white & blue. Blue & white LEDs are more costly in comparison to other LEDs. The color of a Light emitting diodes is decided by the semi-conductor substance, not by the coloring the plastic of the body.

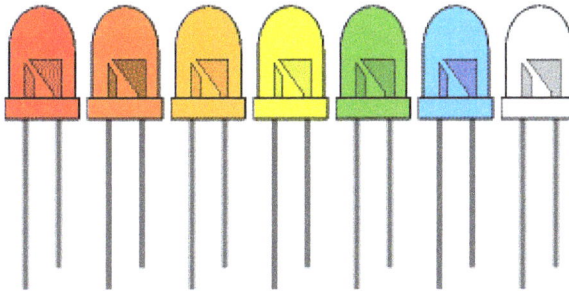

Varactor Diode

Varicap or Varactor diode is that shows the attributes of a variable capacitor. The exhaustion area at the p-n junction behaves as the di-electric and plates of an ordinary capacitor and grounds expansion and contraction by the voltage applied to the varicap diode. This action boosts and reduces the capacitance. The graphic symbol for the varicap diode is shown below. Varactors are employed in fine-tuning circuits and can be employed as high frequency amplifiers.

Anode   Cathode

Even though varactor or varicap diodes can be employed inside several sorts of circuit, they discover applications inside 2 key areas:

- RF filters
- Voltage controlled oscillators, VCOs

Gunn Diode

Gun diodes can also be called as TED (transferred electron devices). These are basically employed in microwave RF devices for frequencies amid 1 and 100 GHz. As microwave RF devices contain Gunn diodes it produces microwave RF signals. The Gunn diode may also be employed for an amplifier and may be acknowledged as TEA (transferred electron amplifier). As Gunn diodes are simple to bring into play, they figure a comparatively low priced technique for producing microwave RF signals.

Gunn diode symbol for circuit diagrams is as follows:

This gun diode is a distinctive component, although it's named as diode, it doesn't comprise of a p-n diode junction. Gunn diodes can be named as diodes because it has 2 electrodes. Gunn diodes are made from a solitary part of n-type semiconductor. The most ordinary materials used are GaAs, InP Indium Phosphide, and gallium Arsenide.

Pin Diode

A PIN diode is a diode with a broad, slightly doped close to inherent semi-conductor area amid an n-type semiconductor and p-type semi-conductor area. The n-type & p-type areas are usually heavily doped for the reason that they are employed for ohmic contacts. The broad inherent area formulates the PIN diode a low rectifier (the common purpose of a diode), but it formulates the PIN diode appropriate for fast switches, photo-detectors, attenuators, and high voltage power electric devices.

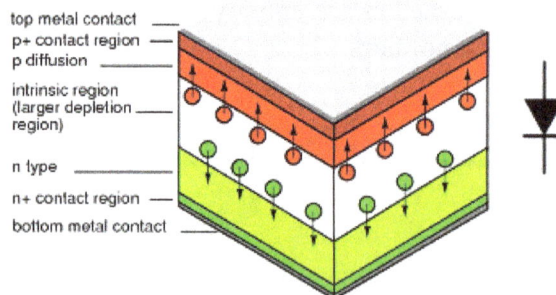

Laser Diode

Laser diode is also named as diode laser or injection diode. It is a semi-conductor appliance that generates coherent radiation (where waves are at similar phase & frequency) in the infrared (IR) or visible spectrum when current is passed via laser diode. Laser diodes are employed in CD (compact disc) players, remote control devices, intrusion detection systems, optical fiber systems and laser printers.

Cathode

Anode

Laser diodes vary from usual lasers, like the He-Ne (helium-neon), gas & ruby types, in a number of ways:

- Laser diodes are small in weight & size.

- Laser diodes are Low in intensity.

- Laser diodes has wide-angle beam>

- Laser diodes require low current, voltage & power.

Photo Diode

Photo diodes are extensively used in various kinds of electronics such as detectors in compact disc players to optical telecommunications systems. Photo diode technology is popular because its trouble-free, inexpensive yet strong configuration. As photodiodes provide dissimilar properties, various photodiode technologies are utilized in a number of areas. There are 4 types of photo diodes:

- PN photodiode

- PIN photodiode

- Schottky photodiode

- Avalanche photodiode

Anode   Cathode

Semiconductor Diode

A diode which is created from semi-conductor material generally- silicon, is known as semiconductor diode. The cathode is negatively charged therefore has a surplus of electrons, is positioned neighboring to the anode, which has intrinsically positive charge, giving a surplus of holes. At this intersection aexhaustion area is created, with no electrons nor holes. Exhaustion area becomes small when a positive voltage is provided to the anode, and current flows; Exhaustion area becomes large when a negative voltage is provides at the anode, thereby stopping current flow.

IMPATT Diode

IMPatt ionization Avalanche Transit Time (Impaat) diode is RF semi-conductor appliance that is utilized for producing microwave radio frequency signals. IMPATT diodes are perfect where small money-making microwave radio supplies are required. These Impatt diodes make brilliant signal supplies for numerous RF microwave devices.

Usually the appliance is utilized in numerous devices including:

- Alarms
- Detectors with RF technology
- Radar

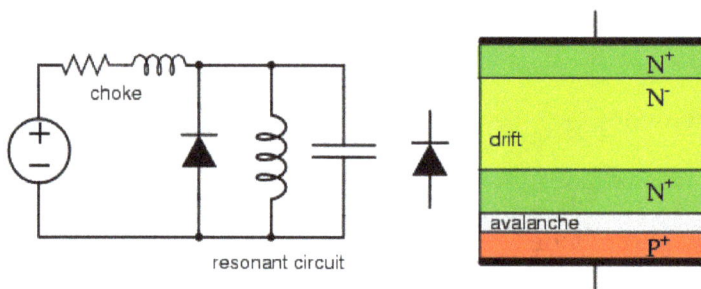

Power Diode

This is a two terminal diode with p-n junction appliance and a p-n junction usually created by permitting circulation & epitaxial developed construction of a power diode. High power diodes are silicon rectifiers that can function at high intersection temperature. Power diodes have superior Voltage, Current & Power managing abilities than usual signal diodes. The power diodes' applications includes battery charging, electric traction, electro plating, power supplies, electro metal processing, welding ups etc.

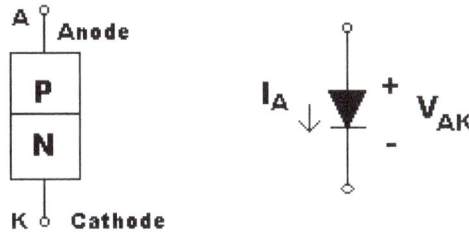

Transistors

Transistor is a semiconductor device that can both conduct and insulate. A transistor can act as a switch and an amplifier. It converts audio waves into electronic waves and resistor, controlling electronic current. Transistors have very long life, smaller in size, can operate on lower voltage supplies for greater safety and required no filament current. The first transistor was fabricated with germanium. A transistor performs the same function as a vacuum tube triode, but using semiconductor junctions instead of heated electrodes in a vacuum chamber. It is the fundamental building block of modern electronic devices and found everywhere in modern electronic systems.

Transistor Basics: A transistor is a three terminal device. Namely,

- Base: This is responsible for activating the transistor.

- Collector: This is the positive lead.

- Emitter: This is the negative lead.

The basic idea behind a transistor is that it lets you control the flow of current through one channel by varying the intensity of a much smaller current that's flowing through a second channel.

The transistor is electronic equipment. It is made through p and n type semiconductor. When a semiconductor is placed in centre between same type semiconductors the arrangement is called transistors. We can say that a transistor is the combination of two diodes it is a connected back to back. A transistor is a device that regulates current or voltage flow and acts as a button or gate for electronic signals. Transistors consist of three layers of a semiconductor device, each capable of moving a current. A semiconductor is a material such like that germanium and silicon that conducts electricity in a "semi-enthusiastic" way. It's anywhere between a genuine conductor such as a copper and an insulator (similar to the plastic wrapped roughly wires).

Transistor Symbol

A diagrammatic form of n-p-n and p-n-p transistor is exposed. In circuit is a connection drawn form is used. The arrow symbol defined the emitter current. In the n-p-n connection we identify electrons flow into the emitter. This means that conservative current flows out of the emitter as an indicated by the outgoing arrow. Equally it can be seen that for p-n-p connection, the conservative current flows into the emitter as exposed by the inward arrow in the figure.

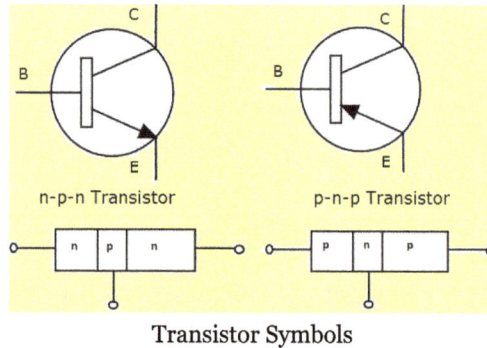

Transistor Symbols

There are so many types of transistors and they each vary in their characteristics and each has their possess advantages and disadvantages. Some types of transistors are used mostly for switching applications. Others can be used for both switching and amplification. Still other transistors are in a specialty group all of their own, such as phototransistors, which react to the amount of light shining on it to produce current flow through it. Below is a list of the different types of transistors; we will go over the characteristics that create them each up.

Bipolar Junction Transistor (BJT)

Bipolar Junction Transistors are transistors which are built up of 3 regions, the base, the collector, and the emitter. Bipolar Junction transistors, different FET transistors, are current-controlled devices. A small current entering in the base region of the transistor causes a much larger current flow from the emitter to the collector region. Bipolar junction transistors come in two major types, NPN and PNP. A NPN transistor is one in which the majority current carrier are electrons. Electron flowing from the emitter to the collector forms the base of the majority of current flow through the transistor. The further types of charge, holes, are a minority. PNP transistors are the opposite. In PNP transistors, the majority current carrier is holes.

Bipolar Junction Transistor pins

Field Effect Transistor

Field Effect Transistors are made up of 3 regions, a gate, a source, and a drain. Different bipolar transistors, FETs are voltage-controlled devices. A voltage placed at the gate controls current flow from the source to the drain of the transistor. Field Effect transistors have a very high input impedance, from several mega ohms ($M\Omega$) of resistance to much, much larger values. This high input impedance causes them to have very little current run through them. (According to ohm's law, current is inversely affected by the value of the impedance of the circuit. If the impedance is high, the current is very low.) So FETs both draw very little current from a circuit's power source.

Field Effect Transistor

Thus, this is ideal because they don't disturb the original circuit power elements to which they are connected to. They won't cause the power source to be loaded down. The drawback of FETs is that they won't provide the same amplification that could be gotten from bipolar transistors. Bipolar transistors are superior in the fact that they provide greater amplification, even though FETs are better in that they cause less loading, are cheaper, and easier to manufacture. Field Effect Transistors come in 2 main types: JFETs and MOSFETs. JFETs and MOSFETs are very similar but MOSFETs have an even higher input impedance values than JFETs. This causes even less loading in a circuit.

Heterojunction Bipolar Transistor (HBT)

AlgaAs/GaAs heterojunction bipolar transistors (HBTs) are used for digital and analog microwave applications with frequencies as high as Ku band. HBTs can supply faster switching speeds than silicon bipolar transistors mostly because of reduced base resistance and collector-to-substrate capacitance. HBT processing requires less demanding lithography than GaAs FETs, therefore, HBTs can priceless to fabricate and can provide better lithographic yield.

This technology can also provide higher breakdown voltages and easier broadband impedance matching than GaAs FETs. In assessment with Si bipolar junction transistors (BJTs), HBTs show better presentation in terms of emitter injection efficiency, base resistance, the base-emitter capacitance, and cutoff frequency. They also present a good linearity, low phase noise and high power-added efficiency. HBTs are used in both profitable and high-reliability applications, such as power amplifiers in mobile telephones and laser drivers.

Darlington Transistor

A Darlington transistor sometimes called as a "Darlington pair" is a transistor circuit that is made from two transistors. Sidney Darlington invented it. It is like a transistor, but it has much higher ability to gain current. The circuit can be made from two discrete transistors or it can be inside an integrated circuit. The hfe parameter with a Darlington transistor is every transistors hfe multiplied mutually. The circuit is helpful in audio amplifiers or in a probe that measures very small current that goes through the water. It is so sensitive that it can pick up the current in the skin. If you connect it to a piece of metal, you can build a touch-sensitive button.

Darlington Transistor

Schottky Transistor

A Schottky transistor is a combination of a transistor and a Schottky diode that prevents the transistor from saturating by diverting the extreme input current. It is also called a Schottky-clamped transistor.

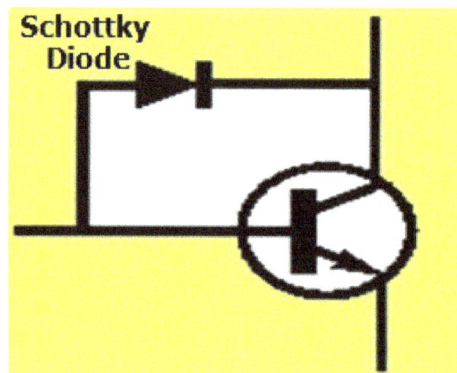

Schottky Transistor

Multiple-emitter Transistor

A multiple-emitter transistor is specialize bipolar transistor frequently used as the inputs of transistor transistor logic (TTL) NAND logic gates. Input signals are applied to the emitters. Collector current stops flowing simply, if all emitters are driven by the logical high voltage, thus performing a NAND logical process using a single transistor. Multiple-emitter transistors replace diodes of DTL and agree to reduction of switching time and power dissipation.

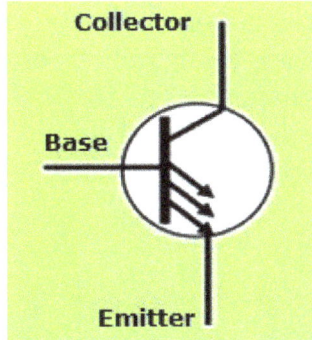

Multiple-Emitter Transistor

Dual Gate MOSFET

One form of MOSFET that is a particularly popular in several RF applications is the dual gate MOSFET. The dual gate MOSFET is used in many RF and other applications where two control gates are required in series. The dual gate MOSFET is fundamentally a form of MOSFET where, two gates are made-up along the length of the channel one after the other.

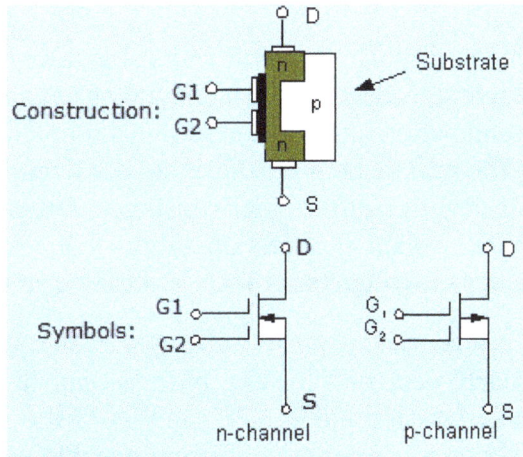

Dual Gate Mosfet

In this way, both gates influence the level of current flowing between the source and drain. In effect, the dual gate MOSFET operation can be considered the same as two MOSFET devices in series. Both gates affect the general MOSFET operation and therefore the output. The dual gate MOSFET can be used in a lot of applications including RF mixers/multipliers, RF amplifiers, amplifiers with gain control and the like.

Junction FET Transistor

The Junction Field Effect Transistor (JUGFET or JFET) has no PN-junctions but in its place has a narrow part of high resistivity semiconductor material forming a "Channel" of either N-type or P-type silicon for the majority carriers to flow through with two ohmic electrical connections at either end normally called the Drain and the Source respectively. There are a two basic configurations of junction field effect transistor, the N-channel JFET and the P-channel JFET. The N-channel JFET's channel is doped with donor impurities meaning that the flow of current through the channel is negative (hence the term N-channel) in the form of electrons.

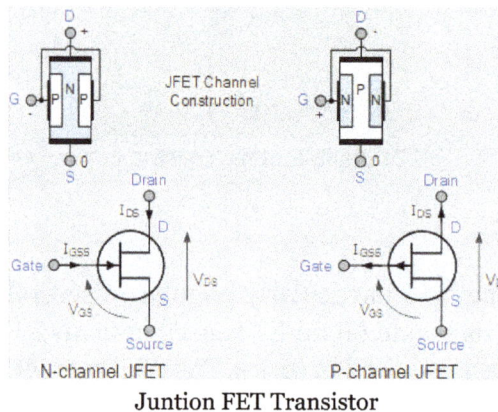

Juntion FET Transistor

Avalanche Transistor

An avalanche transistor is a bipolar junction transistor designed for process in the region of its collector-current/collector-to-emitter voltage characteristics beyond the collector-to-emitter breakdown voltage, called avalanche breakdown region. This region is characterized by the avalanche breakdown, an occurrence similar to Townsend discharge for gases, and negative differential resistance. Operation in the avalanche breakdown region is called avalanche-mode operation: it gives avalanche transistors the capability to switch very high currents with less than a nanosecond rise and fall times (transition times).

Transistors not particularly designed for the purpose can have reasonably consistent avalanche properties; for example 82% of samples of the 15V high-speed switch 2N2369, manufactured over a 12-year period, were capable of generating avalanche breakdown pulses with rise time of 350 ps or less, using a 90V power supply as Jim Williams writes.

Diffusion Transistor

A diffusion transistor is a bipolar junction transistor (BJT) formed by diffusing dopants into a semiconductor substrate. The diffusion process was implemented later than the alloy junction and grown junction processes for making BJTs. Bell Labs developed the first prototype diffusion transistors in 1954. The original diffusion transistors were diffused-base transistors. These transistors still had alloy emitters and sometimes alloy collectors like the earlier alloy-junction transistors. Only the base was diffused into the substrate. Sometimes the substrate produced the collector, but in transistors like Philco's micro alloy diffused transistors the substrate was the bulk of the base.

Transistor Applications

The appropriate application of power semiconductors requires an understanding of their maximum ratings and electrical characteristics, information that is presented within the device data sheet. Good design practice employs data sheet limits and not information obtained from small sample lots. A rating is a maximum or minimum value that sets a limit on device's ability. Act in excess of a rating can result in irreversible degradation or device failure. Maximum ratings signify extreme capabilities of a device. They are not to be used as design circumstances.

A characteristic is a measure of device performance under individual operating conditions expressed by minimum, characteristic, and maximum values, or revealed graphically.

Bipolar Junction Transistor Biasing

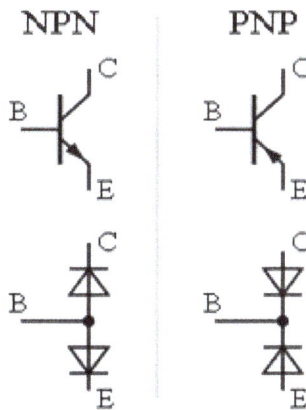

Transistors are the most important semiconductor active devices essential for almost all circuits. They are used as electronic switches, amplifiers etc. in circuits. Transistors may be NPN, PNP, FET, JFET etc. which have different functions in electronic circuits. For the proper working of the circuit, it is necessary to bias the transistor using resistor networks. Operating point is the point on the output characteristics that shows the Collector-Emitter voltage and the Collector current with no input signal. The Operating point is also known as the Bias point or Q-Point (Quiescent point).

Biasing is referred to provide resistors, capacitors or supply voltage etc to provide proper operating characteristics of the transistors. DC biasing is used to obtain DC collector current at a particular collector voltage. The value of this voltage and current are expressed in terms of the Q-Point. In a transistor amplifier configuration, the IC (max) is the maximum current that can flow through the transistor and VCE (max) is the maximum voltage applied across the device. To work the transistor as an amplifier, a load resistor RC must be connected to the collector. Biasing set the DC operating voltage and current to the correct level so that the AC input signal can be properly amplified by the transistor. The correct biasing point is somewhere between the fully ON or fully OFF states of the transistor. This central point is the Q-Point and if the transistor is properly biased, the Q-point will be the central operating point of the transistor. This helps the output current to increase and decrease as the input signal swings through the complete cycle.

For setting the correct Q-Point of the transistor, a collector resistor is used to set the collector current to a constant and steady value without any signal in its base. This steady DC operating point is set by the value of the supply voltage and the value of the base biasing resistor. Base bias resistors are used in all the three transistor configurations like common base, common collector and Common emitter configurations.

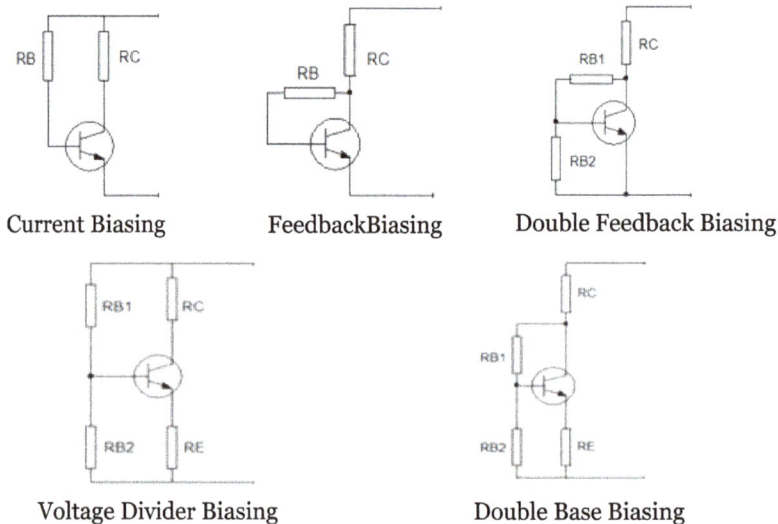

Current Biasing   FeedbackBiasing   Double Feedback Biasing

Voltage Divider Biasing   Double Base Biasing

Modes of Biasing

Following are the different modes of transistor base biasing:

- Current biasing

   As shown in the figure above, two resistors RC and RB are used to set the base bias. These resistors establish the initial operating region of the transistor with a fixed current bias.

The transistor forward biases with a positive base bias voltage through RB. The forward base-Emitter voltage drop is 0.7 volts. Therefore the current through RB is $IB = (V_{cc} - V_{BE}) / I_B$

- Feedback biasing

Figure shows the transistor biasing by the use of a feedback resistor. The base bias is obtained from the collector voltage. The collector feedback ensures that the transistor is always biased in the active region. When the collector current increases, the voltage at the collector drops. This reduces the base drive which in turn reduces the collector current. This feedback configuration is ideal for transistor amplifier designs.

- Double Feedback Biasing

Figure shows how the biasing is achieved using double feedback resistors.

By using two resistors RB1 and RB2 increases the stability with respect to the variations in Beta by increasing the current flow through the base bias resistors. In this configuration, the current in RB1 is equal to 10 % of the collector current.

- Voltage Dividing Biasing

Figure shows the Voltage divider biasing in which two resistors RB1 and RB2 are connected to the base of the transistor forming a voltage divider network. The transistor gets biases by the voltage drop across RB2. This kind of biasing configuration is used widely in amplifier circuits.

- Double Base Biasing

Figure shows a double feedback for stabilization. It uses both Emitter and Collector base feedback to improve the stabilization through controlling the collector current. Resistor values should be selected so as to set the voltage drop across the Emitter resistor 10% of the supply voltage and the current through RB1, 10% of the collector current.

Advantages of Transistor:

- Smaller mechanical sensitivity.

- Lower cost and smaller in size, especially in small-signal circuits.

- Low operating voltages for greater safety, lower costs and tighter clearances.

- Extremely long life.

- No power consumption by a cathode heater.

- Fast switching.

References

- Resistor-types-resistors-fixed-variable-linear-non-linear: electricaltechnology.org, Retrieved 29 July 2018

- Capacitors-introduction-types-applications: engineersgarage.com, Retrieved 21 May 2018

- Types-of-inductors-and-applications: electronicshub.org, Retrieved 18 April 2018

- What-is-a-switch, switch-basics: sparkfun.com, Retrieved 28 March 2018

- What-is-electrical-wire: wisegeek.com, Retrieved 18 June 2018

- Electrical-wires-cables, electrical-materials-products: dfliq.net, Retrieved 13 March 2018

- Transistors-basics-types-baising-modes: elprocus.com, Retrieved 30 April 2018

Electrical Laws and Theorems for Circuit Designing

To develop a comprehension of circuit designing, it is vital to understand the fundamental electrical laws and theorems for circuit designing. This chapter includes various topics central to the understanding of these aspects such as Kirchhoff's current law, Ohm's law, Thévenin's theorem, Norton's theorem, Superposition Theorem, etc.

Kirchhoff's Current Law

Kirchhoff's Current Law helps to solve unknowns when working with electrical circuits. Kirchhoff's.

Current Law with the addition of Kirchhoff's Voltage Law and Ohm's Law will allow for the solution of complex circuits.

Definition that will help in understanding Kirchhoff's Current Law:

Junction - A junction is any point in a circuit where two or more circuit paths come together.

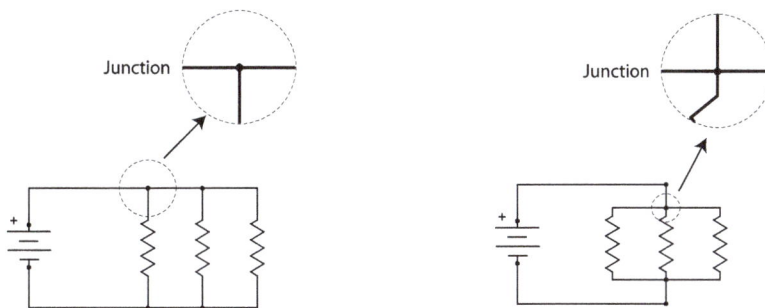

Examples of a Junction

Kirchhoff's Current Law generally states: The algebraic sum of all currents entering (+) and leaving (-) any point (junction) in a circuit must equal zero.

Restated as: The sum of the currents into a junction is equal to the sum of the currents out of that junction.

In figure, there is 3 Amps entering the junction and is split between two paths leaving the junction. The output current, 2 Amps for one path and 1 Amp for the other, has to equal that 3 Amps entering the junction.

Notice that the path for the measurement of the current is through the multimeter. The circuit has to be broken so that the current can go through the multimeter.

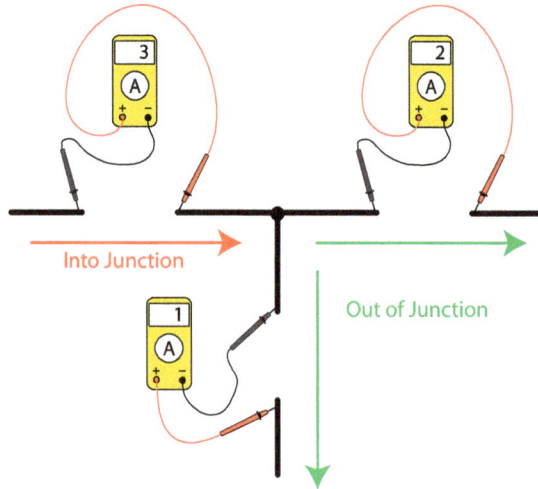

Kirchhoff's Current Law expressed algebraically:

The algebraic sum of all currents entering (+) and leaving (-) any point (junction) in a circuit must equal zero.

$$I(in\ 1) + I(in\ 2) + I(in...) + I(out\ 1) + I(out\ 2) + I(out\ ...) = 0$$

Example: 3 A + (-2 A) + (-1 A) = 0

The example above is the math equivalence for the junction in the above figure. The negative sign indicates direction only (leaving a junction) and does not represent a negative current. This formula, although correct can be somewhat confusing. By rewriting the formula it can be made easier to understand.

Kirchhoff's Current Law Rewritten

The sum of the currents into a junction is equal to the sum of the currents out of that junction.

I(in 1) + I(in 2) + I(in...) = I(out 1) + I(out 2) + I(out ...)

Example: 3 A = 2 A + 1 A

Converting to this equation removes the need to work with the negatives and for some people this equation is easier to remember and work with.

For a complete circuit this equation can be written using Current Total (I_t) on one side of the equation and all paths or branches on the other side of the equation.

Application of Kirchhoff's Current Law in a Series Circuit

Remember that by definition a Series circuit has only one path for current. That means that the Total Current leaving the Voltage source has only one path back to the Voltage source. This further means that at any point along the path of a Series circuit the same amount of current will be flowing.

An equation that shows how Kirchhoff's Current Law applies to Series circuits is:

$$It = I_{R1} = I_{R2} = I_{Rn...}$$

I_{R1} stands for the Current (I) through R_1. Likewise for the other Currents in the equation above.

The implications of this equation is that if you can solve for any Current (I_t, I_{R1}, I_{R2}, ...) in a series circuit you will know what the Current is for all of the circuit. This idea is sometimes stated as "Current is common in a series circuit."

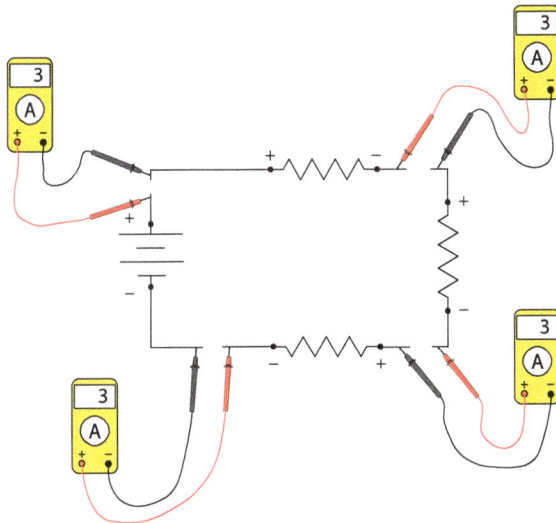

In figure above, notice that all of the multimeter measurement readings are the same and all are positive. The red probe of each multimeter is electrically closest to the positive terminal of the Voltage source, and the black probe is closest to the negative terminal of the Voltage source.

If the leads were reversed for any of the multimeter a negative reading would be made. This indicates that the Current is flowing through the meter in the opposite direction; this does not change the fact that the Current in the circuit is "common" and the formula stays the same.

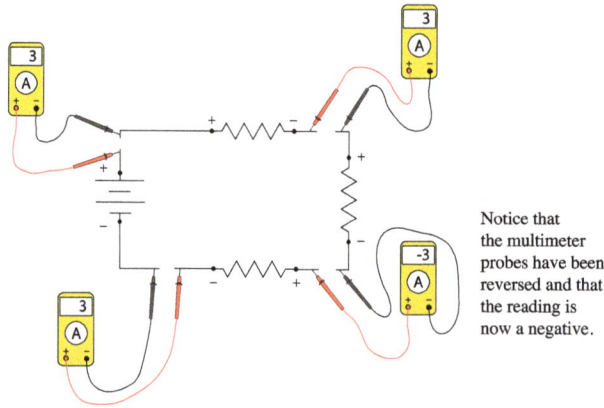

Notice that
the multimeter
probes have been
reversed and that
the reading is
now a negative.

Solving for Current in a Series Circuit

In an earlier you will find 3 equations for solving for Current with Ohm's/Watt's Law.

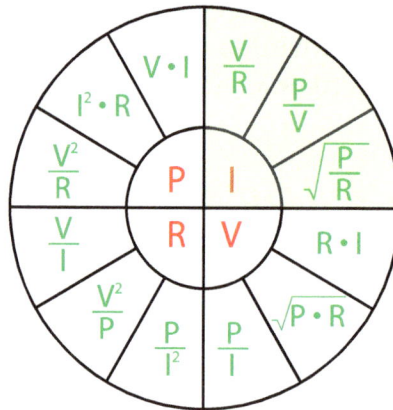

If given a Series Circuit to solve for Total I (I_t):

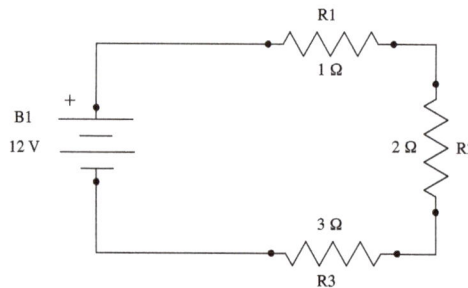

One way to get the Total Current in the circuit above is to solve for the Total Resistance Rt using the equation Rt = R_1 + R_2 + R_n..., which would give a value of 6 Ω for the circuit above. Once the value for V_t (12 ÷ V) and the value for R_t (6 Ω) is known, the equation V_t ÷ R_t = It can be used. 12V ÷ 6 Ω = 2A. So in the circuit above the Total Current flowing through the circuit is 2 Amps.

Using Kirchhoff's Current Law for a Series circuit $I_t = I_{R1} = I_{R2} = I_{Rn}$. The current through all of the resistors are known. The Current through $R_1 = I_t$ or 2 Amps, the Current through $R_2 = I_t$ or 2 Amps, and finally the Current through $R_3 = I_t$ or 2 Amps.

Is there another way to solve for the Currents in the above circuit? Yes, in the Kirchhoff's Voltage Law topic there was an equation to find the Voltage Drop for each known resistance in a series circuit.

Let's use R_1 as an example, once the Voltage Drop is known for a resistance (V_{R1}) and the resistance value is known R_1, then I_{R1} can be calculated using $V_{R1} \div R_1 = I_{R1}$. Remember $I_{R1} = It = I_{R2} = I_{Rn}$.

$$\frac{R_1}{Rt} \times V_s = V_{R1} \qquad \text{So,} \qquad \frac{1\Omega}{6\Omega} \times 12V = 2V$$

Solving for the Voltage Drop across R_1

$$\frac{V_{R1}}{R_1} = I_{R1} \qquad \text{So,} \qquad \frac{2V}{1\Omega} = 2\,A$$

Solving for the Current through R_1.

Again, in a series circuit, if the current through one electrical component (Resistor in this circuit) is known, then the current through all the electrical components (Resistors in this circuit) is known.

Application of Kirchhoff's Current Law in a Parallel Circuit

Note that the Current going through R_1 (3A), R_2 (4A), and R_3 (5A) add up to the total Current leaving the Voltage Source B1 (12A). Before it is shown how to calculate for these individual Currents, an understanding of the current flowing through the circuit will be shown. Figure below shows the first branch (R1) of the above parallel circuit, 12 A enters the junction and is split into two paths, 3 A going to R_1 and 9 A going toward R_2 and R_3. This complies with Kirchhoff's Current Law, 12 = 9 + 3.

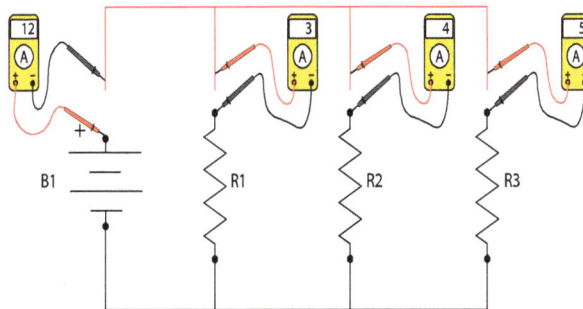

$$12\,A = 3\,A + 4\,A + 5\,A$$

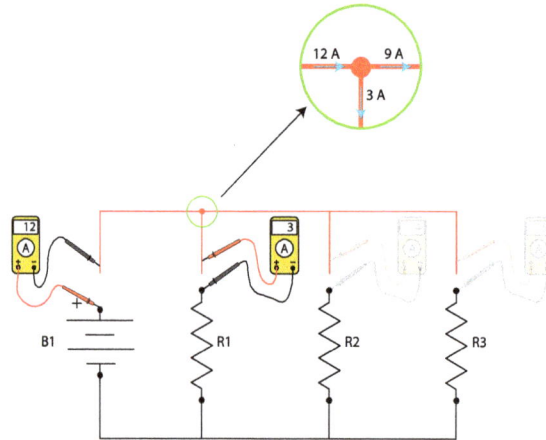

Figure below shows the second branch (R_2) of the parallel circuit, 9 A enters the junction and is split into two paths, 4 A going to R_2 and 5A going toward R_3. This complies with Kirchhoff's Current Law, 9 = 4 + 5.

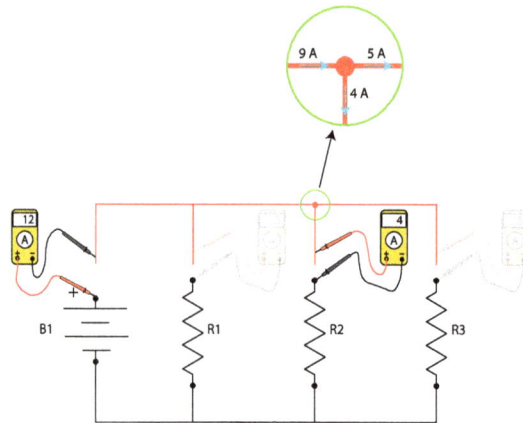

Solving for Current in a Parallel Circuit

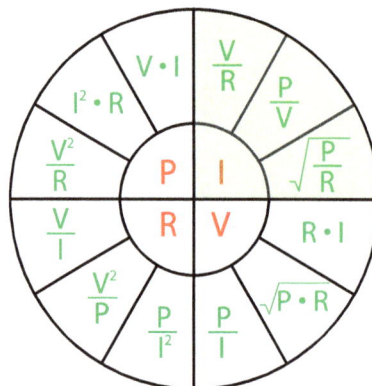

If given a Parallel Circuit to solve for Total I (I_t):

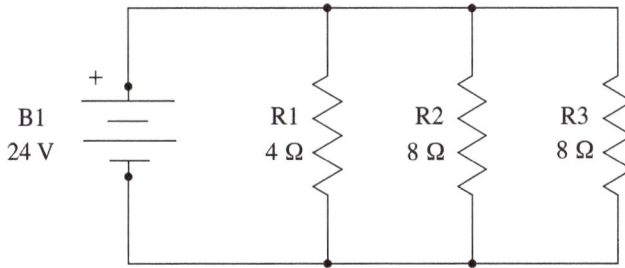

One way to get the Total Current for the circuit above is to solve for the Total Resistance (R_t) using the equation $R_t = 1 \div ((1 \div R_1) + (1 \div R_2) + (1 \div R_3))$, which would give a value of 2 Ω. Once the value for Vt (24 V) and the value for Rt (2 Ω) is known, the equation $V_t \div R_t = I_t$ can be used; 24 V ÷ 2 Ω = 12 A. In the circuit above the Total Current flowing through the circuit is 12 Amps.

Another way to solve for the Total Current is to solve for all the individual branch Currents and then add them together. This is due to the additive nature of Kirchhoff's Current Law for parallel circuits ($I_t = I_{R1} + I_{R2} + I_{Rn}...$).

Kirchhoff's Voltage Law for a parallel circuit can help solve for Current. Kirchhoff's Voltage Law for a parallel circuit indicates that the "Voltage is common" is a parallel circuit. ($V_t = V_{R1} = V_{R2} = V_{R3}$).

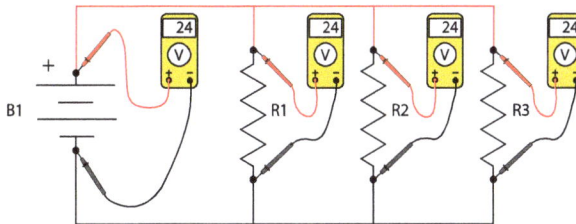

To solve for any of the Currents going through the resistors in the parallel circuit in figure above, two electrical perimeters must be known. In this case Kirchhoff's Voltage Law provided the Voltage across each resistor (V_{R1}, V_{R2}, and V_{R3}). The Resistive values where given (R_1, R_2, and R_3). Using Ohm's Law, the Current can be calculated using V ÷ R = I.

Solving for the Current through R_1: $\dfrac{V_{R1}}{R_1} = I_{R1} \quad \rightarrow \quad \dfrac{24\,V}{4\Omega} = 6A$

Solving for the Current through R_2: $\dfrac{V_{R2}}{R_2} = I_{R2} \quad \rightarrow \quad \dfrac{24\,V}{8\Omega} = 3\,A$

Solving for the Current through R_3: $\dfrac{V_{R3}}{R_3} = I_{R3} \quad \rightarrow \quad \dfrac{24\,V}{8\Omega} = 3\,A$

Once the Current is known for all of the branches in a parallel circuit, the total current for the circuit can be found be adding all of the individual Currents together. Kirchhoff's Current Law for a parallel circuit can be stated as $I_t = I_{R_1} + I_{R_2} + I_{Rn}$

Solving for the total current: $I_t = I_{R_1} + I_{R_2} + I_{R_3}$ 12 A = 6 A + 3 A + 3 A

Kirchoff's Voltage Law

Gustav Kirchhoff's Voltage Law is the second of his fundamental laws we can use for circuit analysis. His voltage law states that for a closed loop series path the algebraic sum of all the voltages around any closed loop in a circuit is equal to zero. This is because a circuit loop is a closed conducting path so no energy is lost.

In other words the algebraic sum of ALL the potential differences around the loop must be equal to zero as: $\Sigma V = 0$. Note here that the term "algebraic sum" means to take into account the polarities and signs of the sources and voltage drops around the loop.

This idea by Kirchhoff is commonly known as the Conservation of Energy, as moving around a closed loop, or circuit, you will end up back to where you started in the circuit and therefore back to the same initial potential with no loss of voltage around the loop. Hence any voltage drops around the loop must be equal to any voltage sources met along the way.

So when applying Kirchhoff's voltage law to a specific circuit element, it is important that we pay special attention to the algebraic signs, (+ and -) of the voltage drops across elements and the emf's of sources otherwise our calculations may be wrong.

But before we look more closely at Kirchhoff's voltage law (KVL) lets first understand the voltage drop across a single element such as a resistor.

A Single Circuit Element

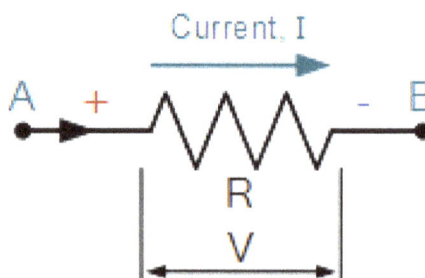

For this simple example we will assume that the current, I is in the same direction as the flow of positive charge, that is conventional current flow.

Here the flow of current through the resistor is from point A to point B, that is from positive terminal to a negative terminal. Thus as we are travelling in the same direction as current flow, there will be a *fall* in potential across the resistive element giving rise to a -IR voltage drop across it.

If the flow of current was in the opposite direction from point B to point A, then there would be a *rise* in potential across the resistive element as we are moving from a -potential to a + potential giving us a +I*R voltage drop.

Thus to apply Kirchhoff's voltage law correctly to a circuit, we must first understand the direction of the polarity and as we can see, the sign of the voltage drop across the resistive element will depend on the direction of the current flowing through it. As a general rule, you will loose potential in the same direction of current across an element and gain potential as you move in the direction of an emf source.

The direction of current flow around a closed circuit can be assumed to be either clockwise or anticlockwise and either one can be chosen. If the direction chosen is different from the actual direction of current flow, the result will still be correct and valid but will result in the algebraic answer having a minus sign.

To understand this idea a little more, lets look at a single circuit loop to see if Kirchhoff's Voltage Law holds true.

A Single Circuit Loop

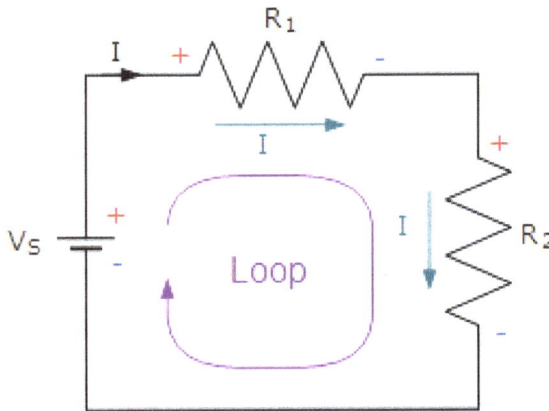

Kirchhoff's voltage law states that the algebraic sum of the potential differences in any loop must be equal to zero as: $\Sigma V = 0$. Since the two resistors, R_1 and R_2 are wired together in a series connection, they are both part of the same loop so the same current must flow through each resistor.

Thus the voltage drop across resistor, $R_1 = I*R_1$ and the voltage drop across resistor, $R_2 = I*R_2$ giving by KVL:

$$V_s + \left(-IR_1\right) + \left(-IR_2\right) = 0$$
$$\therefore V_s = IR_1 + IR_2$$
$$V_s = I\left(R_1 + R_2\right)$$
$$V_s = IR_T$$

We can see that applying Kirchhoff's Voltage Law to this single closed loop produces the formula for the equivalent or total resistance in the series circuit and we can expand on this to find the values of the voltage drops around the loop.

$$R_T = R_1 + R_2$$
$$I = \frac{V_s}{R_T} = \frac{V_s}{R_1 + R_2}$$
$$V_{R1} = IR_1 = V_s\left(\frac{R_1}{R_1 + R_2}\right)$$
$$V_{R2} = IR_2 = V_s\left(\frac{R_1}{R_1 + R_2}\right)$$

Kirchhoff's Voltage Law Example

Three resistors of values: 10 ohms, 20 ohms and 30 ohms, respectively are connected in series across a 12 volt battery supply. Calculate:

a) The total resistance,

b) The circuit current,

c) The current through each resistor,

d) The voltage drop across each resistor,

e) Verify that Kirchhoff's voltage law, KVL holds true.

a) Total Resistance (R_T)

$$R_T = R_1 + R_2 + R_3 = 10\Omega + 20\Omega + 30\Omega = 60\Omega$$

Then the total circuit resistance RT is equal to 60Ω

b) Circuit Current (I)

$$I = \frac{V_s}{R_T} = \frac{12}{60} = 0.2A$$

Thus the total circuit current I is equal to 0.2 amperes or 200mA

c) Current Through Each Resistor

The resistors are wired together in series; they are all part of the same loop and therefore each experience the same amount of current. Thus:

$$I_{R1} = I_{R2} = I_{R3} = I_{SERIES} = 0.2 \text{ amperes}$$

d) Voltage Drop Across Each Resistor

$$V_{R1} = I \times R_1 = 0.2 \times 10 = 2 \text{volts}$$
$$V_{R2} = I \times R_2 = 0.2 \times 20 = 4 \text{volts}$$
$$V_{R3} = I \times R_3 = 0.2 \times 30 = 6 \text{volts}$$

e) Verify Kirchhoff's Voltage Law

$$V_s + \left(-I_{R1}\right) + \left(-I_{R2}\right) + \left(-I_{R3}\right) = 0$$
$$12 + \left(-0.20 \times 10\right) + \left(-0.2 \times 20\right) + \left(0.2 \times 30\right) = 0$$
$$12 + \left(-2\right) + \left(-4\right) + \left(-6\right) = 0$$
$$\therefore 12 - 2 - 4 - 6 = 0$$

Thus Kirchhoff's voltage law holds true as the individual voltage drops around the closed loop add up to the total.

Kirchhoff's Circuit Loop

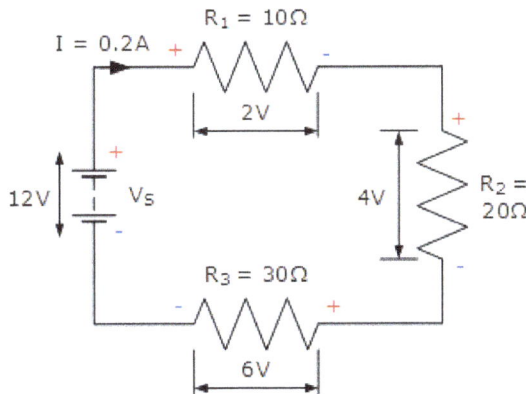

We have seen here that Kirchhoff's voltage law, KVL is Kirchhoff's second law and states that the algebraic sum of all the voltage drops, as you go around a closed circuit from some fixed point and return back to the same point, and taking polarity into account, is always zero. That is $\Sigma V = 0$

The theory behind Kirchhoff's second law is also known as the law of conservation of voltage, and this is particularly useful for us when dealing with series circuits, as series circuits also act as voltage dividers and the voltage divider circuit is an important application of many series circuits.

Ohm's Law

When an electric potential difference (V) is applied across a conductor as shown in the figure below, some current (I) flows through it. The flow of current is opposed by the resistance of the conductor and circuit. The relation between voltage, current and resistance is explained by the ohm's law.

Ohm's laws state that the current through any two points of the conductor is directly proportional to the potential difference applied across the conductor, provided physical conditions i.e. temperature, etc. do not change. It is measured in (Ω) ohm.

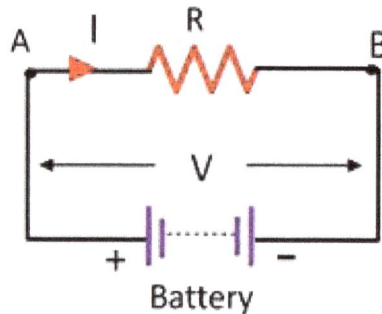

Battery

Mathematically it is expressed as;

$$I \alpha V$$

$$\frac{V}{I} = constant$$

$$\frac{V_1}{I_1} = \frac{V_2}{I_2} = ... = \frac{V_n}{I_n} = constant$$

In other words, Ohm's law can also be stated as;

The ratio of the potential difference across the end point of the conductor to the current flowing between them is always constant, but the physical conditions of the conductor i.e. temperature, etc. remain same.

This constant is also called the resistance (R) of the conductor (or circuit),

$$\frac{V}{I} = R$$

It can be written as,

$$I = \frac{V}{R}$$

In a circuit, when current flows through a resistor, the potential difference across the resistor is known as voltage drops across it, i.e., V = IR.

Limitations of Ohm's Law

- Ohm's law is not applicable in unilateral networks. Unilateral networks allow the current to flow in one direction. Such types of network consist elements like a diode, transistor, etc.

- It is not applicable for the non-linear network. In the nonlinear network, the parameter of the network is varied with the voltage and current. Their parameter likes resistance, inductance, capacitance and frequency, etc., not remain constant with the times. So ohms law is not applicable to the nonlinear network.

Ohm's law are used for finding the resistance of the circuit and also for knowing the voltage and current of the circuit.

Thevenin's Theorem

Thevenin's Theorem states that – any complicated network across its load terminals can be substituted by a voltage source with one resistance in series. This theorem helps in the study of the variation of current in a particular branch when the resistance of the branch is varied while the remaining network remains the same for example designing of electronic circuits.

A more general statement of Thevenin's Theorem is that any linear active network consisting of independent or dependent voltage and current source and the network elements can be replaced by an equivalent circuit having a voltage source in series with a resistance, that voltage source being the open circuited voltage across the open circuited load terminals and the resistance being the internal resistance of the source.

In other words, the current flowing through a resistor connected across any two terminals of a network by an equivalent circuit having a voltage source Eth in series with a resistor R_{th}. Where E_{th} is the open circuit voltage between the required two terminals called the Thevenin voltage and the R_{th} is the equivalent resistance of the network as seen from the two terminal with all other sources replaced by their internal resistances called Thevenin resistance.

Explanation of Thevenin's Theorem

The Thevenin's statement is explained with the help of a circuit shown below:

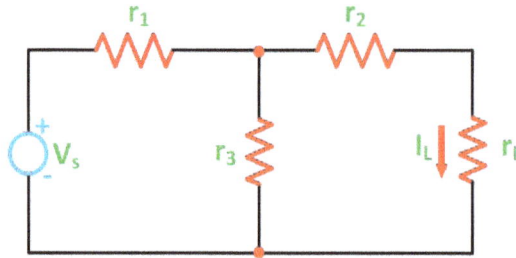

Let us consider a simple DC circuit as shown in the figure above, where we have to find the load current I_L by the Thevenin's theorem. In order to find the equivalent voltage source, r_L is removed from the circuit as shown in the figure below and V_{oc} or V_{TH} is calculated,

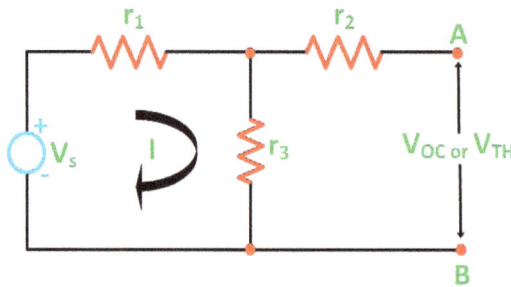

$$V_{OC} = I\, r_3 = \frac{V_S}{r_1 + r_3} r_3$$

Now, to find the internal resistance of the network (Thevenin's resistance or equivalent resistance) in series with the open circuit voltage V_{OC}, also known as Thevenin's voltage V_{TH}, the voltage source is removed or we can say it is deactivated by a short circuit (as the source does not have any internal resistance) as shown in the figure below,

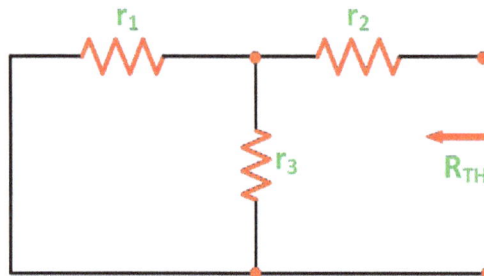

$$R_{TH} = r_2 + \frac{r_1 r_3}{r_1 + r_3}$$

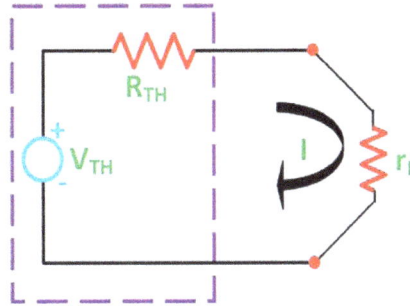

Equivalent Circuit of Thevenin's Theorem

As per Thevenin's Statement, the load current is determined by the circuit shown above and the equivalent Thevenin's circuit is obtained.

The Load current I_L is given as,

$$I_L = \frac{V_{TH}}{R_{TH} + r_L}$$

Where,

V_{TH} is the Thevenin's equivalent voltage. It is an open circuit voltage across the terminal AB known as load terminal.

R_{TH} is the Thevenin's equivalent resistance, as seen from the load terminals where all the sources are replaced by their internal impedance.

r_L is the load resistance.

Steps for Solving Thevenin's Theorem

Step 1: First of all remove the load resistance r_L of the given circuit.

Step 2: Replace all the impedance source by their internal resistance.

Step 3: If sources are ideal then short circuit the voltage source and open the current source.

Step 4: Now find the equivalent resistance at the load terminals know as Thevenin's Resistance (R_{TH}).

Step 5: Draw the Thevenin's equivalent circuit by connecting the load resistance and after that determine the desired response.

This theorem is possibly the most extensively used networks theorem. It is applicable where it is desired to determine the current through or voltage across any one element in a network. The Thevenin's Theorem is an easy way to solve the complicated network.

Norton's Theorem

Norton's Theorem states that – A linear active network consisting of independent or dependent voltage source and current sources and the various circuit elements can be substituted by an equivalent circuit consisting of a current source in parallel with a resistance. The current source being the short-circuited current across the load terminal and the resistance being the internal resistance of the source network.

The Norton's theorems reduce the networks equivalent to the circuit having one current source, parallel resistance and load. Norton's theorem is the converse of Thevenin's Theorem. It consists of the equivalent current source instead of an equivalent voltage source as in Thevenin's theorem. The determination of internal resistance of the source network is identical in both the theorems.

In the final stage that is in the equivalent circuit, the current is placed in parallel to the internal resistance in Norton's Theorem whereas in Thevenin's Theorem the equivalent voltage source is placed in series with the internal resistance.

Explanation of Norton's Theorem

To understand Norton's Theorem in detail, let us consider a circuit diagram given below,

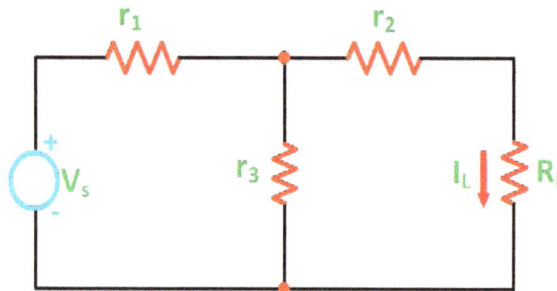

In order to find the current through the load resistance IL as shown in the circuit diagram above, the load resistance has to be short-circuited as shown in the diagram below,

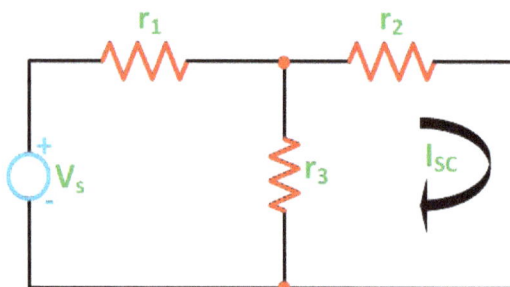

Now, the value of current I flowing in the circuit is found out by the equation,

$$I = \frac{V_S}{r_1 + \dfrac{r_2 r_3}{r_2 + r_3}}$$

And the short-circuit current I_{SC} is given by the equation shown below,

$$I_{SC} = I \frac{r_3}{r_3 + r_2}$$

Now the short circuit is removed, and the independent source is deactivated as shown in the circuit diagram below and the value of the internal resistance is calculated by,

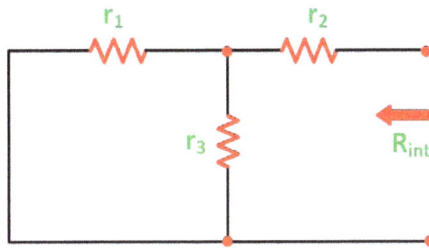

$$R_{int} = r_2 + \frac{r_1 r_3}{r_1 + r_3}$$

As per the Norton's Theorem, the equivalent source circuit would contain a current source in parallel to the internal resistance, the current source being the short-circuited current across the shorted terminals of the load resistor. The Norton's Equivalent circuit is represented as,

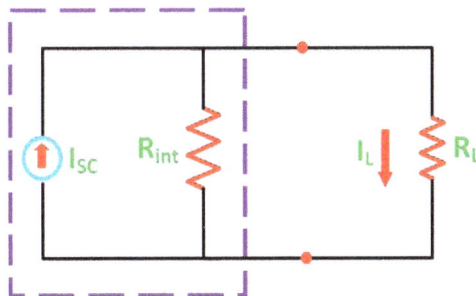

Finally the load current I_L calculated by the equation shown below,

$$I_L = I_{SC} \frac{R_{int}}{R_{int} + r_L}$$

Where,

- I_L is the load current

- I_{sc} is the short circuit current

- R_{int} is the internal resistance of the circuit

- R_L is the load resistance of the circuit

Steps for Solving a Network Utilizing Norton's Theorem

Step 1: Remove the load resistance of the circuit.

Step 2: Find the internal resistance R_{int} of the source network by deactivating the constant sources.

Step 3: Now short the load terminals and find the short circuit current I_{SC} flowing through the shorted load terminals using conventional network analysis methods.

Step 4: Norton's equivalent circuit is drawn by keeping the internal resistance R_{int} in parallel with the short circuit current I_{SC}.

Step 5 – Reconnect the load resistance RL of the circuit across the load terminals and find the current through it known as load current I_L.

Superposition Theorem

In a linear circuit with several sources the voltage and current responses in any branch is the algebraic sum of the voltage and current responses due to each source acting independently with all other sources replaced by their internal impedance.

OR

In any linear circuit containing multiple independent sources, the current or voltage at any point in the network may be calculated as algebraic sum of the individual contributions of each source acting alone.

The Process of using Superposition Theorem on a Circuit

To solve a circuit with the help of Superposition theorem follow the following steps:

- First of all make sure the circuit is a linear circuit; or a circuit where Ohm's law implies, because Superposition theorem is applicable only to linear circuits and responses.

- Replace all the voltage and current sources on the circuit except for one of them.

While replacing a Voltage source or Current Source replace it with their internal resistance or impedance. If the Source is an Ideal source or internal impedance is not given then replace a Voltage source with a short; so as to maintain a 0 V potential difference between two terminals of the voltage source. And replace a Current source with an Open; so as to maintain a 0 Amps Current between two terminals of the current source.

- Determine the branch responses or voltage drop and current on every branch simply by using KCL, KVL or Ohm's Law.

- Repeat step 2 and 3 for every source the circuit has.

- Now algebraically add the responses due to each source on a branch to find the response on the branch due to the combined effect of all the sources.

The superposition theorem is not applicable for the power, as power is directly proportional to the square of the current which is not a linear function.

Steps

1. Select any one source and short all other voltage sources and open all current sources if internal impedance is not known. If known replace them by their impedance.

2. Find out the current or voltage across the required element, due to the source under consideration.

3. Repeat the above steps for all other sources.

4. Add all the individual effects produced by individual sources to obtain the total current in or across the voltage element.

Example

Find I in circuit shown below, where only the current source is kept in the circuit. The 5V is zeroed out yielding a 0V source, or a short. The 9V is zeroed out, making it a short also.

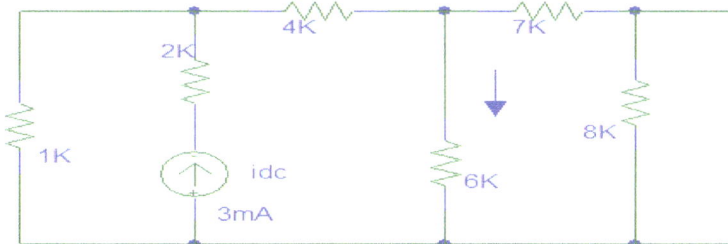

Note that the 8K resistor is shorted out (that is, 8K in parallel with 0 yields 0.

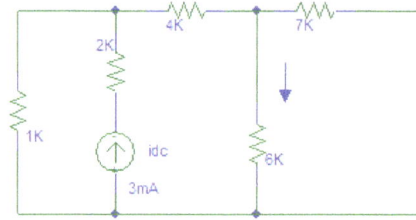

Note that the 3mA flowing up through the 2K will split left and right at the top.

Part of it will flow through the 1K and part of it will flow through the 4K. Let's use the label "I4" for the current flowing right through the 4K resistor. If we combine the parallel 6K and 7K (6K||7K = 3.2K) and then add the series 4K, the total resistance on the right is 7.2K. Now we can use a current divider to find that I4 = [1K / (1K + 7.2K)] * 3mA = 0.37mA. Note that the 2K does not enter into this computation because the entire 3mA flows through it. The 3mA does not split until it gets to the junction at the top of that branch.

Now that we know I4, we can then split it again through the 6K and the 7K. Ix = [7K/(6K+7K)]*I4 = 0.20 mA.

Example

Using the superposition theorem, determine the voltage drop and current across the resistor 3.3K as shown in figure below:

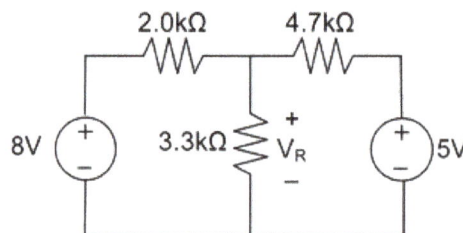

Solution

Step 1: Remove the 8V power supply from the original circuit, such that the new circuit becomes as the following and then measure voltage across resistor.

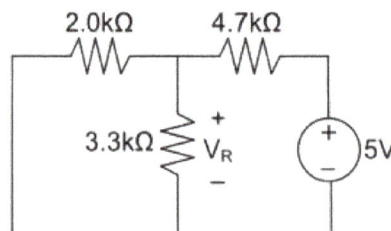

Here 3.3K and 2K are in parallel; therefore resultant resistance will be 1.245K.

Using voltage divider rule voltage across 1.245K will be,

$\quad$ V1= [1.245/(1.245+4.7)]*5 = 1.047V

Step 2: Remove the 5V power supply from the original circuit such that the new circuit becomes as the following and then measure voltage across resistor.

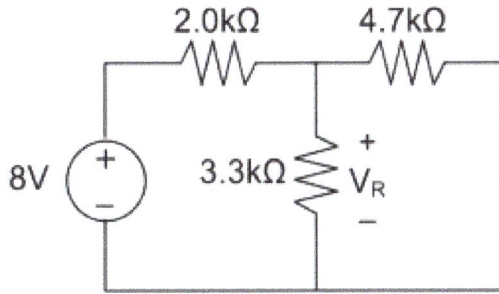

Here 3.3K and 4.7K are in parallel, therefore resultant resistance will be 1.938K.

Using voltage divider rule voltage across 1.938K will be,

$\quad$ V2= [1.938/(1.938+2)]*8 = 3.9377V

Therefore voltage drop across 3.3K resistor is V1+V2 = 1.047+3.9377=4.9847

Reciprocity Theorem

Reciprocity theorem is one of the DC network analysis and AC network analysis technique and deals with the relationship between impressed source in a part of the circuit and it's response at some other part of the circuit.

The Reciprocal Theorem states that: In any bilateral linear circuits; If a source of EMF "V" acting in a branch (let "A") of the circuit produces the current "I" in another branch (let "B") of the circuit. Then when the EMF "V" acts in the second branch "B", it will produce the same current "I" in the first branch "A".

In another words, the supply voltage "V" and a 0-ohm ammeter reading or current "I" are mutually transferable in any bilateral linear circuits. The ratio between the Voltage and the Current that are mutually transferable is called the transfer resistance.

The Reciprocity theorem holds true in bilateral linear circuits, or on linear circuits which contains only bilateral components; this property of the bilateral linear circuits is called their reciprocal property and the circuits in which the Reciprocity theorem holds true are called reciprocal circuits.

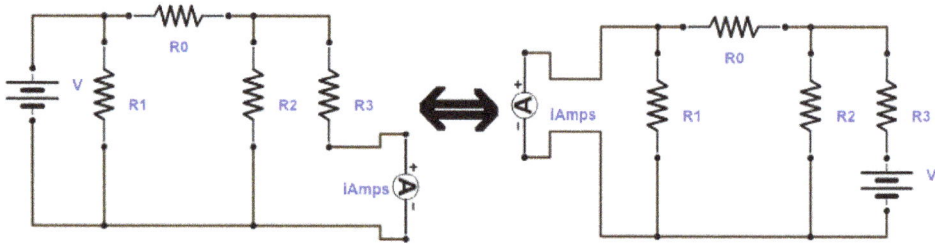

Reciprocity Theorem

Reciprocity theorem can be applied to solve many DC and AC electrical network easily and efficiently and it also has especial applications in electromagnetism and antenna electronics.

Millman's Theorem

In Millman's Theorem, the circuit is re-drawn as a parallel network of branches, each branch containing a resistor or series battery/resistor combination. Millman's Theorem is applicable only to those circuits which can be re-drawn accordingly. Here is an example used for the method analysis.

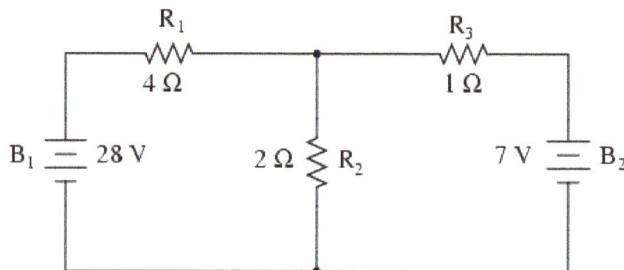

And here is that same circuit, re-drawn for the sake of applying Millman's Theorem:

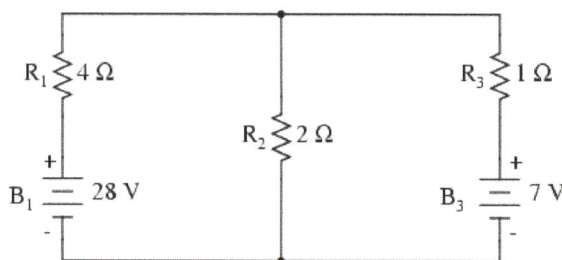

By considering the supply voltage within each branch and the resistance within each branch, Millman's Theorem will tell us the voltage across all branches. Please note that the labeled the battery in the rightmost branch as "B$_3$" to clearly denote it as being in the third branch, even though there is no "B$_2$" in the circuit.

Millman's Theorem is nothing more than a long equation, applied to any circuit drawn as a set of parallel-connected branches, each branch with its own voltage source and series resistance:

Millman's Theorem Equation

$$\frac{\dfrac{E_{B1}}{R_1} + \dfrac{E_{B2}}{R_2} + \dfrac{E_{B3}}{R_3}}{\dfrac{1}{R_1} + \dfrac{1}{R_2} + \dfrac{1}{R_2}} = \text{Voltage across all branches}$$

Substituting actual voltage and resistance figures from our example circuit for the variable terms of this equation, we get the following expression:

$$\frac{\dfrac{28V}{4\Omega} + \dfrac{0V}{2\Omega} + \dfrac{7V}{1\Omega}}{\dfrac{1}{4\Omega} + \dfrac{1}{2\Omega} + \dfrac{1}{1\Omega}} = 8V$$

The final answer of 8 volts is the voltage seen across all parallel branches, like this:

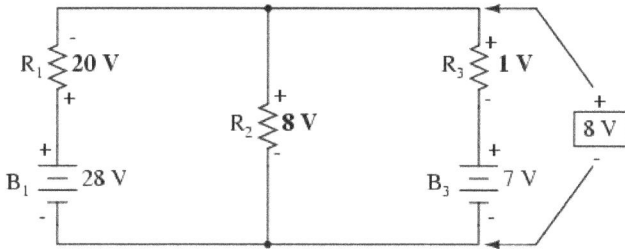

The polarity of all voltages in Millman's Theorem are referenced to the same point. In the example circuit above, I used the bottom wire of the parallel circuit as my reference point, and so the voltages within each branch (28 for the R1 branch, 0 for the R$_2$ branch, and 7 for the R$_3$ branch) were inserted into the equation as positive numbers. Likewise, when the answer came out to 8 volts (positive), this meant that the top wire of the circuit was positive with respect to the bottom wire (the original point of reference). If both batteries had been connected backwards (negative ends up and positive ends down), the voltage for branch 1 would have been entered into the equation as a -28 volts, the voltage for branch 3 as -7 volts, and the resulting answer of -8 volts would have told us that the top wire was negative with respect to the bottom wire.

To solve for resistor voltage drops, the Millman voltage (across the parallel network) must be compared against the voltage source within each branch, using the principle of voltages adding in series to determine the magnitude and polarity of voltage across each resistor:

$$E_{R_1} = 8V\text{-}28V = \text{-}20V\,(\text{negative on top})$$
$$E_{R_2} = 8V\text{-}0V = 8V\,(\text{positive on top})$$
$$E_{R_3} = 8V\text{-}7V = 1V\,(\text{positive on top})$$

To solve for branch currents, each resistor voltage drop can be divided by its respective resistance (I=E/R):

$$I_{R_1} = \frac{20V}{4\Omega} = 5A$$
$$I_{R_2} = \frac{8V}{2\Omega} = 4A$$
$$I_{R_3} = \frac{1V}{1\Omega} = 1A$$

The direction of current through each resistor is determined by the polarity across each resistor, not by the polarity across each battery, as current can be forced backwards through a battery, as is the case with B3 in the example circuit. This is important to keep in mind, since Millman's Theorem doesn't provide as direct an indication of "wrong" current direction as does the Branch Current or Mesh Current methods. You must pay close attention to the polarities of resistor voltage drops as given by Kirchhoff's Voltage Law, determining direction of currents from that.

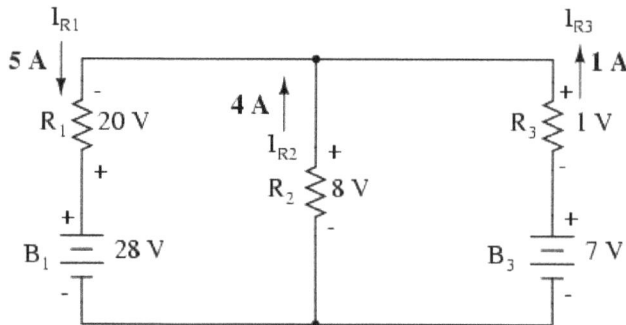

Millman's Theorem is very convenient for determining the voltage across a set of parallel branches, where there are enough voltage sources present to preclude solution via regular series-parallel reduction method. It also is easy in the sense that it doesn't require the use of simultaneous equations. However, it is limited in that it only applied to circuits which can be re-drawn to fit this form. It cannot be used, for example, to solve an unbalanced bridge circuit. And, even in cases where Millman's Theorem can be applied, the solution of individual resistor voltage drops can be a bit daunting to some, the Millman's Theorem equation only providing a single figure for branch voltage.

Compensation Theorem

Compensation Theorem states that in a linear time invariant network when the resistance (R) of an uncoupled branch, carrying a current (I), is changed by (ΔR). The currents in all the branches would change and can be obtained by assuming that an ideal voltage source of (VC) has been connected such that VC= I (ΔR) in series with (R + ΔR) when all other sources in the network are replaced by their internal resistances.

In Compensation Theorem, the source voltage (V_c) opposes the original current. In simple words compensation theorem can be stated as – the resistance of any network can be replaced by a voltage source, having the same voltage as the voltage drop across the resistance which is replaced.

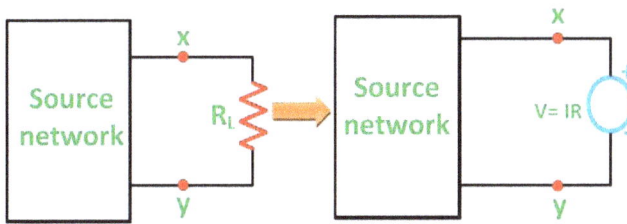

Explanation

Let us assume a load RL be connected to a DC source network whose Thevenin's equivalent gives Vo as the Thevenin's voltage and RTH as the Thevenin's resistance as shown in the figure below:

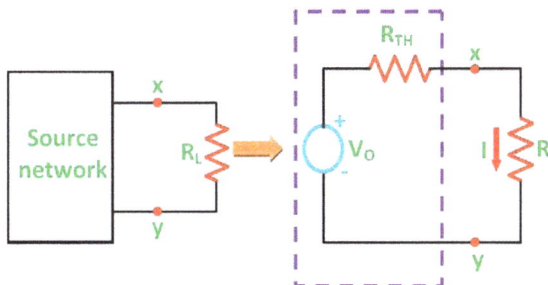

Here,

$$I' = \frac{V_o}{R_{TH} + R_L}$$

Let the load resistance RL be changed to (RL + ΔRL). Since the rest of the circuit remains unchanged, the Thevenin's equivalent network remains the same as shown in the circuit diagram below:

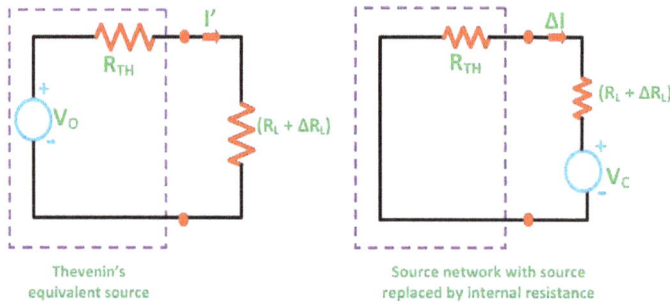

Thevenin's
equivalent source

Source network with source
replaced by internal resistance

Here,

$$I' = \frac{V_o}{R_{TH} + (R_L + \Delta R_L)}$$

The change of current being termed as ΔI

Therefore,

$$\Delta I = I' - I$$

Putting the value of I' and I from the equation $\left(I' = \dfrac{V_o}{R_{TH} + R_L} \right)$ and

$\left(I' = \dfrac{V_o}{R_{TH} + (R_L + \Delta R_L)} \right)$ in the equation ($\Delta I = I' - I$) we will get the following equation

$$\Delta I = \frac{V_o}{R_{TH} + (R_L + \Delta R_L)} - \frac{V_o}{R_{TH} + R_L}$$

$$\Delta I = \frac{V_o \{ R_{TH} + R_L - (R_{TH} + R_L + \Delta R_L) \}}{(R_{TH} + R_L + \Delta R_L)(R_{TH} + R_L)}$$

$$\Delta I = -\left[\frac{V_o}{R_{TH} + R_L} \right] \frac{\Delta R_L}{R_{TH} + R_L + \Delta R_L}$$

Now, putting the value of I from the equation $\left(I' = \dfrac{V_o}{R_{TH} + R_L} \right)$ in the equation

($\Delta I = -\left[\dfrac{V_o}{R_{TH} + R_L} \right] \dfrac{\Delta R_L}{R_{TH} + R_L + \Delta R_L}$), we will get the following equation

$$\Delta I = -\frac{I \Delta R_L}{R_{TH} + R_L + \Delta R_L}$$

As we know, VC = I Δ RL and is known as compensating voltage.

Therefore, the equation ($\Delta I = -\dfrac{I\Delta R_L}{R_{TH} + R_L + \Delta R_L}$) becomes

$$\Delta I = \dfrac{-V_C}{R_{TH} + R_L + \Delta R_L}$$

Hence, Compensation Theorem tells that with the change of branch resistance, branch currents changes and the change is equivalent to an ideal compensating voltage source in series with the branch opposing the original current, all other sources in the network being replaced by their internal resistances.

Substitution Theorem

As name implies, the main concept of this theorem which is based upon substitution of one element by another equivalent element. Substitution theorem gives us some special insights in circuit behavior. This theorem is also used to prove several other theorems.

Substitution theorem states that if an element in a network is replaced by a voltage source whose voltage at any instant of time is equals to the voltage across the element in the previous network then the initial condition in the rest of the network will be unaltered or alternately if an element in a network is replaced by a current source whose current at any instant of time is equal to the current through the element in the previous network then the initial condition in the rest of the network will be un-altered.

Let us take a circuit as shown in figure,

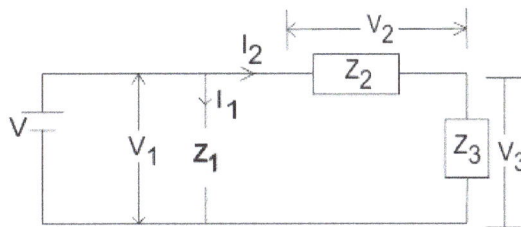

Let, V is supplied voltage and Z_1, Z_2 and Z_3 is different circuit impedances. V_1, V_2 and V_3 are the voltages across Z_1, Z_2 and Z_3 impedance respectively and I is the supplied current whose I_1 part is flowing through the Z_1 impedance whereas I_2 part is flowing through the Z_2 and Z_3 impedance.

Now if we replace Z_3 impedance with V_3 voltage source as shown in figure or with I_2 current source as shown in figure then according to Substitution Theorem all initial condition through other impedances and source will remain unchanged.

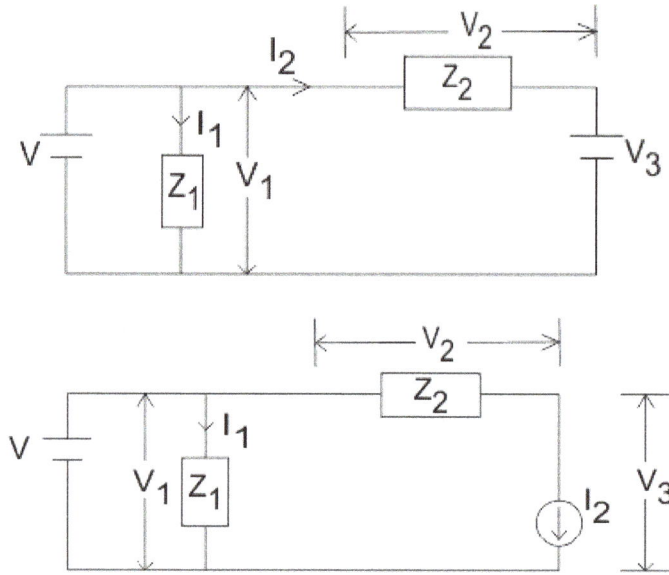

i.e. - current through source will be I, voltage across Z_1 impedance will be V_1, current through Z_2 will be I_2 etc.

Example

For more efficient and clear understanding let us go through a simple practical example: Let us take a circuit as shown in figure below:

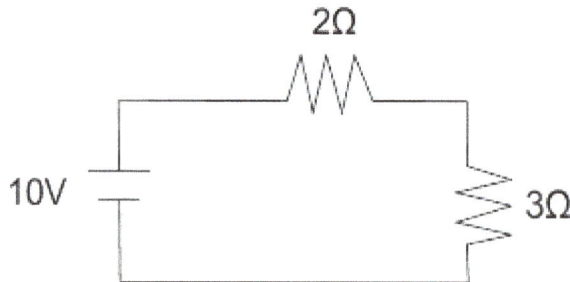

As per voltage division rule voltage across 3Ω and 2Ω resistance are,

$$V_{3\Omega} = \frac{10 \times 3}{3+2} = 6V$$

$$V_{2\Omega} = \frac{10 \times 2}{3+2} = 4V$$

Current through the circuit, $I = \dfrac{10}{3+2} = 2A$

If we replace the 3Ω resistance with a voltage source of 6V as shown in figure below,

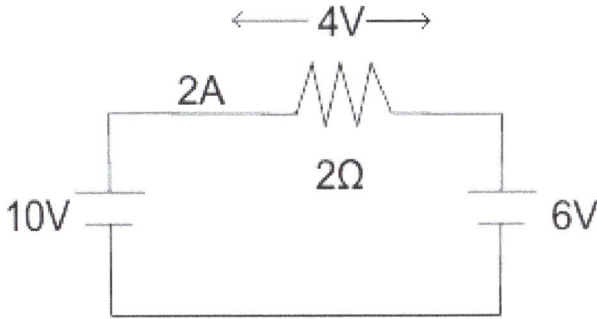

According to Ohm's law the voltage across 2Ω resistance and current through the circuit is,

$$V_{2\Omega} = 10 - 6 = 4V$$

$$I = \frac{10-6}{2} 2A$$

Alternately if we replace 3Ω resistance with a current source of 2A as shown in figure below,

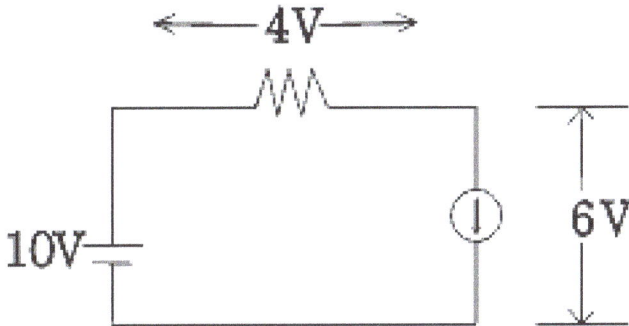

Voltage across 2Ω is $V_{2\Omega}$ = 10 - 3× 2 = 4 V and voltage across 2A current source is V_{2A} = 10 - 4 = 6 V. We can see the voltage across 2Ω resistance and current through the circuit is unaltered i.e all initial condition of the circuit is intact.

Maximum Power Transfer Theorem

The Maximum Power Transfer Theorem is another useful circuit analysis method to ensure that the maximum amount of power will be dissipated in the load resistance when the value of the load resistance is exactly equal to the resistance of the power source. The relationship between the load impedance and the internal impedance of the energy source will give the power in the load.

Thevenins Equivalent Circuit

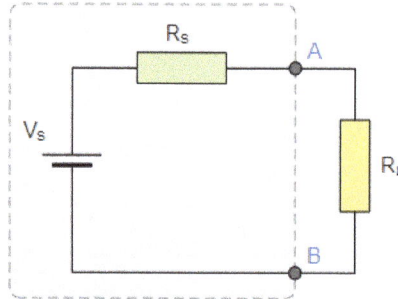

In our Thevenin equivalent circuit above, the maximum power transfer theorem states that "*the maximum amount of power will be dissipated in the load resistance if it is equal in value to the Thevenin or Norton source resistance of the network supplying the power*".

In other words, the load resistance resulting in greatest power dissipation must be equal in value to the equivalent Thevenin source resistance, then $R_L = R_S$ but if the load resistance is lower or higher in value than the Thevenin source resistance of the network, its dissipated power will be less than maximum.

For example, find the value of the load resistance, R_L that will give the maximum power transfer in the following circuit.

Maximum Power Transfer Example

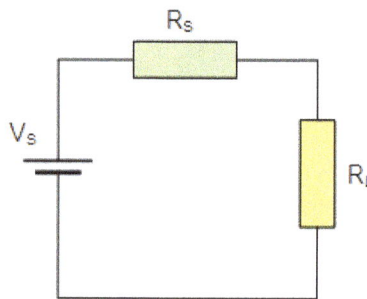

Where,

$R_S = 25\Omega$

R_L is variable between 0 – 100Ω

$V_S = 100v$

Then by using the following Ohm's Law equations,

$$I = \frac{V_s}{R_S + R_L} \quad \text{and} \quad P = I^2 R_L$$

We can now complete the following table to determine the current and power in the circuit for different values of load resistance.

Table of Current Against Power

R_L (Ω)	I (amps)	P (watts)	RL (Ω)	I (amps)	P (watts)
0	4.0	0	25	2.0	**100**
5	3.3	55	30	1.8	97
10	2.8	78	40	1.5	94
15	2.5	93	60	1.2	83
20	2.2	97	100	0.8	64

Using the data from the table above, we can plot a graph of load resistance, R_L against power, P for different values of load resistance. Also notice that power is zero for an open-circuit (zero current condition) and also for a short-circuit (zero voltage condition).

Graph of Power Against Load Resistance

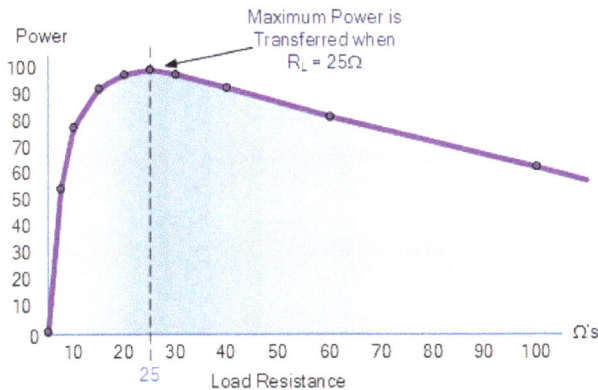

From the above table and graph we can see that the Maximum Power Transfer occurs in the load when the load resistance, R_L is equal in value to the source resistance, RS that is: $R_S = R_L = 25\Omega$. This is called a "matched condition" and as a general rule, maximum power is transferred from an active device such as a power supply or battery to an external device when the impedance of the external device exactly matches the impedance of the source.

One good example of impedance matching is between an audio amplifier and a loudspeaker. The output impedance, Z_{OUT} of the amplifier may be given as between 4Ω and 8Ω, while the nominal input impedance, Z_{IN} of the loudspeaker may be given as 8Ω only.

Then if the 8Ω speaker is attached to the amplifiers output, the amplifier will see the speaker as an 8Ω load. Connecting two 8Ω speakers in parallel is equivalent to the amplifier driving one 4Ω speaker and both configurations are within the output specifications of the amplifier.

Improper impedance matching can lead to excessive power loss and heat dissipation. But how could you impedance match an amplifier and loudspeaker which have very different impedances. Well, there are loudspeaker impedance matching transformers available that can change impedances from 4Ω to 8Ω, or to 16Ω's to allow impedance matching of many loudspeakers connected together in various combinations such as in PA (public address) systems.

Transformer Impedance Matching

One very useful application of impedance matching in order to provide maximum power transfer between the source and the load is in the output stages of amplifier circuits. Signal transformers are used to match the loudspeakers higher or lower impedance value to the amplifiers output impedance to obtain maximum sound power output. These audio signal transformers are called "matching transformers" and couple the load to the amplifiers output as shown below.

Transformer Impedance Matching

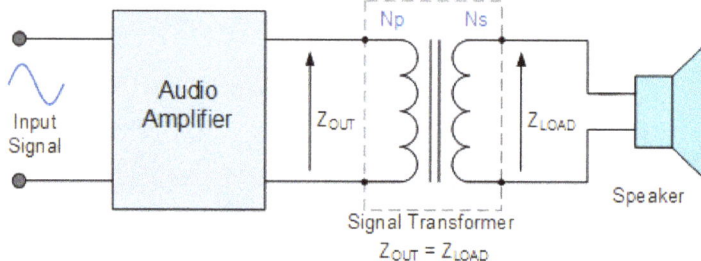

The maximum power transfer can be obtained even if the output impedance is not the same as the load impedance. This can be done using a suitable "turns ratio" on the transformer with the corresponding ratio of load impedance, Z_{LOAD} to output impedance, Z_{OUT} matches that of the ratio of the transformers primary turns to secondary turns as a resistance on one side of the transformer becomes a different value on the other.

If the load impedance, Z_{LOAD} is purely resistive and the source impedance is purely resistive, Z_{OUT} then the equation for finding the maximum power transfer is given as:

$$Z_{out} = \left(\frac{N_P}{N_S}\right)^2 Z_{load}$$

Where: N_P is the number of primary turns and N_S the number of secondary turns on the transformer. Then by varying the value of the transformers turns ratio the output impedance can be "matched" to the source impedance to achieve maximum power transfer.

Maximum Power Transfer Example 2

If an 8Ω loudspeaker is to be connected to an amplifier with an output impedance of 1000Ω, calculate the turns ratio of the matching transformer required to provide maximum power transfer of the audio signal. Assume the amplifier source impedance is Z_1, the load impedance is Z_2 with the turns ratio given as N.

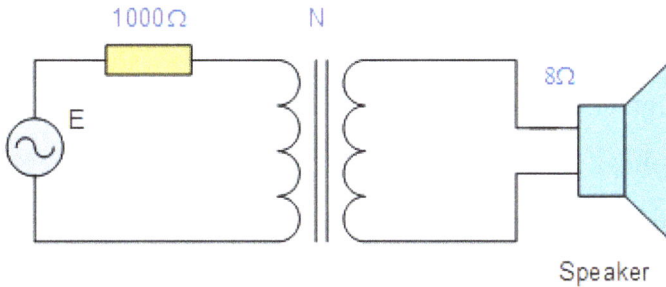

$$Z_1 = N^2 Z_2 \therefore N = \sqrt{\frac{Z_1}{Z_2}}$$

therefore,

$$N = \sqrt{\frac{Z_1}{Z_2}} = \sqrt{\frac{1000}{8}} = 11.2:1$$

Generally, small high frequency audio transformers used in low power amplifier circuits are nearly always regarded as ideal for simplicity, so any losses can be ignored.

Tellegen Theorem

Tellegen's Theorem states that the summation of power delivered is zero for each branch of any electrical network at any instant of time. It is mainly applicable for designing the filters in signal processing. It is also used in complex operation systems for regulating the stability. It is mostly used in the chemical and biological system and for finding the dynamic behaviour of the physical network.

Tellegen's theorem is independent of the network elements. Thus, it is applicable for any lump system that has linear, active, passive and time-variant elements. Also, the theorem is convenient for the network which follows Kirchhoff's current law and Kirchhoff's voltage law.

Explanation of tellegen's theorem

Tellegen's Theorem can also be stated in another word as, in any linear, nonlinear, passive, active, time variant or time invariant network the summation of power (instantaneous or complex power of sources) is zero.

Thus, for the Kth branch, this theorem states that

$$\sum_{K=1}^{n} v_K i_K = 0$$

Where,

n is the number of branches

v_K is the voltage in the branch

i_K is the current flowing through the branch

Let

$$i_{pq} = i_K$$

Equation ($i_{pq} = i_K$) shows the K_{th} branch through current

vK is the voltage drop in branch K and is given as

$$K = v_p - v_q$$

Where v_p and v_q are the respective node voltage at p and q nodes.

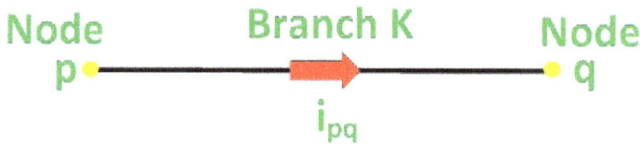

We have,

$$v_K i_{pq} = \left(v_p - v_q\right) i_{pq} = v_K i_K$$

Also

$$v_K i_K = \left(v_p - v_q\right) i_{pq}$$

Obviously

$$i_{qp} = -i_{pq}$$

Summing the above two equations ($v_K i_{pq} = (v_p - v_q) i_{pq} = v_K i_K$) and ($v_K i_K = (v_p - v_q) i_{pq}$), we get

$$2v_K i_K = (v_p - v_q) i_{pq} + (v_p - v_q) i_{pq} \quad \text{or}$$

$$v_K i_K = \frac{1}{2} \left[(v_p - v_q) i_{pq} + (v_p - v_q) i_{pq} \right]$$

Such equations can be written for every branch of the network.

Assuming n branches the equation will be

$$\sum_{K=1}^{n} v_K i_K = \frac{1}{2} \sum_{p=1}^{n} \sum_{q=1}^{n} (v_p - v_q) i_{pq}$$

$$\sum_{K=1}^{n} v_K i_K = \frac{1}{2} \sum_{p=1}^{n} v_p \left[\sum_{q=1}^{n} i_{pq} \right] - \frac{1}{2} \sum_{q=1}^{n} v_q \left[\sum_{p=1}^{n} i_{pq} \right]$$

However, according to the Kirchhoff's current law (KCL), the algebraic sum of currents at each node is equal to zero.

Therefore,

$$\sum_{p=1}^{n} i_{pq} = 0 \quad and \quad \sum_{q=1}^{n} i_{qp} = 0$$

Thus, from the above equation ($\sum_{p=1}^{n} i_{pq} = 0 \quad and \quad \sum_{q=1}^{n} i_{qp} = 0$) finally we obtain

$$\sum_{K=1}^{n} v_K i_K = 0$$

Thus, it has been observed that the sum of power delivered to a closed network is zero. This proves that the Tellegen's Theorem and also proves the conservation of power in any electrical network. It is also evident that the sum of power delivered to the network by an independent source is equal to the sum of power absorbed by all passive elements of the network.

Steps for Solving Networks Using Tellegen's Theorem

Step 1: The following steps are given below to solve any electrical network by Tellegen's Theorem.

Step 2: In order to justify this theorem in an electrical network, the first step is to find the branch voltage drops.

Step 3: Find the corresponding branch currents using conventional analysis methods.

Step 4: Tellegen's Theorem can then be justified by summing the products of all branch voltages and currents.

For example, if a network having some branches "b" then,

$$\sum_{b=1}^{b} v_b i_b = 0$$

Now if the set of voltages and currents is taken, corresponding the two different instants of time, t_1 and t_2, the Tellegen's Theorem is also applicable where we get the equation as shown below,

$$\sum_{b=1}^{b} v_b(t_1) i_b(t_2) = \sum_{b=1}^{b} v_b(t_2) i_b(t_1) = 0$$

Application of Tellegen's Theorem

The various applications of the Tellegen's theorem are as follows:

- It is used in the digital signal processing system for designing of filters.

- In the area of the biological and chemical process.

- In topology and structure of reaction network analysis.

- The theorem is used in chemical plants and oil industries to determine the stability of any complex systems.

Prototype Filter

Prototype filters are electronic filter designs that are used as a template to produce a modified filter design for a particular application. They are an example of a non dimensionalised design from which the desired filter can be scaled or transformed. They are most often seen in regard to electronic filters and especially linear analogue passive filters. However, in principle, the method can be applied to any kind of linear filter or signal processing, including mechanical, acoustic and optical filters.

Operand Isolation

In electronic low power digital synchronous circuit design, operand isolation is a technique for minimizing the energy overhead associated with redundant operations by selectively blocking the propagation of switching activity through the circuit. This technique isolates sections of the circuit (operation) from "seeing" changes on their inputs (operands) unless they are expected to respond to them. This is usually done using latches at the inputs of the circuit. The latches become transparent only when the result of the operation is going to be used. One can also use multiplexers or simple AND gates instead of latches.

Operand isolation reduces dynamic power dissipation. When the enable is inactive, the datapath inputs are disabled so that unnecessary switching power is not wasted in the datapath.

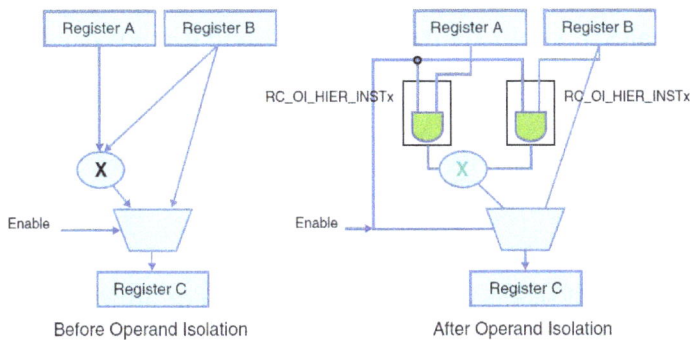

Before Operand Isolation   After Operand Isolation

In the digital system shown as before operand isolation, register C uses the result of the multiplier when the enable is on. When the enable is off, register C uses only the result of register B, but the multiplier continues its computations. Because the multiplier dissipates the most power, the total amount of power wasted is quite significant.

One solution to this problem is to shut down (isolate) the function unit (operand) when its results are not used, as shown in after operand isolation. The synthesis engine inserts AND gates at the inputs of the multiplier and uses the enable logic of the multiplier to gate the signal transitions. As a result, no dynamic power is dissipated when the result of the multiplier is not needed.

For a sample of designs operand isolation provided less than 5% dynamic power savings with little or no impact on any other aspects of the design.

References

- Kirchhoffs-voltage-law: electronics-tutorials.ws, Retrieved 15 July 2018

- What-is-thevenins-theorem: circuitglobe.com, Retrieved 30 May 2018

- Reciprocity-theorem: electronicspani.com, Retrieved 27 June 2018

- What-is-compensation-theorem: circuitglobe.com, Retrieved 17 April 2018

- What-is-tellegens-theorem: circuitglobe.com, Retrieved 31 March 2018

- Operand-isolation-735297: semanticscholar.org, Retrieved 29 March 2018

4

Circuit Analysis Techniques

Circuit analysis is the procedure of determining the currents and voltages across electrical components of a network. This chapter has been carefully written to provide an easy understanding of the varied facets of circuit analysis techniques, such as network analysis, mesh analysis, super mesh transform, nodal analysis, distributed element model, large-signal model, symbolic circuit analysis, etc.

Network Analysis

A network, in the context of electronics, is a collection of interconnected components. Network analysis is the process of finding the voltages across, and the currents through, every component in the network. There are many different techniques for calculating these values. However, for the most part, the applied technique assumes that the components of the network are all linear. The methods described in this topic are only applicable to *linear* network analysis, except where explicitly stated.

Component	A device with two or more terminals into which, or out of which, current may flow.
Node	A point at which terminals of more than two components are joined. A conductor with a substantially zero resistance is considered to be a node for the purpose of analysis.
Branch	The component(s) joining two nodes.
Mesh	A group of branches within a network joined so as to form a complete loop such that there is no other loop inside it .
Port	Two terminals where the current into one is identical to the current out of the other.
Circuit	A current from one terminal of a generator, through load component(s) and back into the other terminal. A circuit is, in this sense, a one-port network and is a trivial case to analyse. If there is any connection to any other circuits then a non-trivial network has been formed and at least two ports must exist. Often, "circuit" and "network" are used interchangeably, but many analysts reserve "network" to mean an idealised model consisting of ideal components.
Transfer function	The relationship of the currents and/or voltages between two ports. Most often, an input port and an output port are discussed and the transfer function is described as gain or attenuation.

	For a two-terminal component (i.e. one-port component), the current and voltage are taken as the input and output and the transfer function will have units of impedance or admittance (it is usually a matter of arbitrary convenience whether voltage or current is considered the input). A three (or more) terminal component effectively has two (or more) ports and the transfer function cannot be expressed as a single impedance. The usual approach is to express the transfer function as a matrix of parameters. These parameters can be impedances, but there is a large number of other approaches.
Component transfer function	

Equivalent Circuits

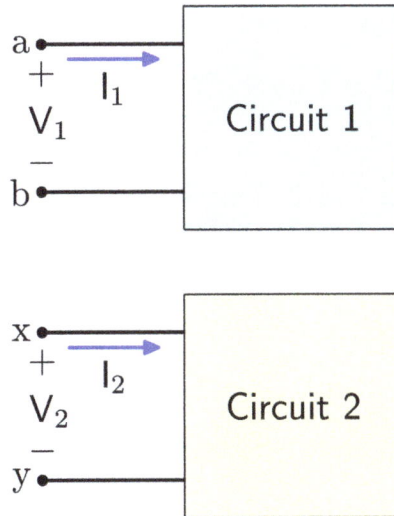

A useful procedure in network analysis is to simplify the network by reducing the number of components. This can be done by replacing the actual components with other notional components that have the same effect. A particular technique might directly reduce the number of components, for instance by combining impedances in series. On the other hand, it might merely change the form into one in which the components can be reduced in a later operation. For instance, one might transform a voltage generator into a current generator using Norton's theorem in order to be able to later combine the internal resistance of the generator with a parallel impedance load.

A resistive circuit is a circuit containing only resistors, ideal current sources, and ideal voltage sources. If the sources are constant (DC) sources, the result is a DC circuit. Analysis of a circuit consists of solving for the voltages and currents present in the circuit. The solution principles outlined here also apply to phasor analysis of AC circuits.

Two circuits are said to be equivalent with respect to a pair of terminals if the voltage across the terminals and current through the terminals for one network have the same relationship as the voltage and current at the terminals of the other network.

If $V_2 = V_1$ implies $I_2 = I_1$ for all (real) values of V_1, then with respect to terminals ab and xy, circuit 1 and circuit 2 are equivalent.

The above is a sufficient definition for a one-port network. For more than one port, then it must be defined that the currents and voltages between all pairs of corresponding ports must bear the same relationship. For instance, star and delta networks are effectively three port networks and hence require three simultaneous equations to fully specify their equivalence.

Impedances in Series and in Parallel

Any two terminal network of impedances can eventually be reduced to a single impedance by successive applications of impedances in series or impedances in parallel.

Impedances in series: $Z_{eq} = Z_1 + Z_2 + \cdots + Z_n$.

Impedances in parallel: $\dfrac{1}{Z_{eq}} = \dfrac{1}{Z_1} + \dfrac{1}{Z_2} + \cdots + \dfrac{1}{Z_n}$

The above simplified for only two impedances in parallel: $Z_{eq} = \dfrac{Z_1 Z_2}{Z_1 + Z_2}$.

Delta-wye Transformation

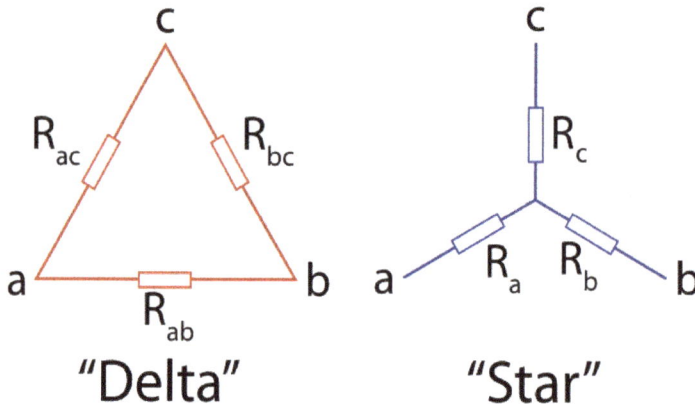

"Delta"   "Star"

A network of impedances with more than two terminals cannot be reduced to a single impedance equivalent circuit. An n-terminal network can, at best, be reduced to n impedances (at worst $^nC_2$). For a three terminal network, the three impedances can be expressed as a three node delta (Δ) network or four node star (Y) network. These two networks are equivalent and the transformations between them are given below. A general network with an arbitrary number of nodes cannot be reduced to the minimum number of impedances using only series and parallel combinations. In general, Y-Δ and Δ-Y transformations must also be used. For some networks the extension of Y-Δ to star-polygon transformations may also be required.

For equivalence, the impedances between any pair of terminals must be the same for both networks, resulting in a set of three simultaneous equations. The equations below are expressed as resistances but apply equally to the general case with impedances.

Delta-to-star Transformation Equations

$$R_a = \frac{R_{ac}R_{ab}}{R_{ac}+R_{ab}+R_{bc}}$$

$$R_b = \frac{R_{ab}R_{bc}}{R_{ac}+R_{ab}+R_{bc}}$$

$$R_c = \frac{R_{bc}R_{ac}}{R_{ac}+R_{ab}+R_{bc}}$$

Star-to-delta Transformation Equations

$$R_{ac} = \frac{R_aR_b+R_bR_c+R_cR_a}{R_b}$$

$$R_{ab} = \frac{R_aR_b+R_bR_c+R_cR_a}{R_c}$$

$$R_{bc} = \frac{R_aR_b+R_bR_c+R_cR_a}{R_a}$$

General Form of Network Node Elimination

The star-to-delta and series-resistor transformations are special cases of the general resistor network node elimination algorithm. Any node connected by N resistors $(R_1 \dots R_N)$ to nodes 1 .. N can be replaced by $\binom{N}{2}$ resistors interconnecting the remaining N nodes. The resistance between any two nodes x and y is given by:

$$R_{xy} = R_xR_y\sum_{i=1}^{N}\frac{1}{R_i}$$

For a star-to-delta ($N = 3$) this reduces to:

$$R_{ab} = R_aR_b(\frac{1}{R_a}+\frac{1}{R_b}+\frac{1}{R_c}) = \frac{R_aR_b(R_aR_b+R_aR_c+R_bR_c)}{R_aR_bR_c} = \frac{R_aR_b+R_bR_c+R_cR_a}{R_c}$$

For a series reduction ($N = 2$) this reduces to:

$$R_{ab} = R_aR_b(\frac{1}{R_a}+\frac{1}{R_b}) = \frac{R_aR_b(R_a+R_b)}{R_aR_b} = R_a+R_b$$

For a Dangling Resistor ($N = 1$) it Results in the Elimination of the Resistor because $\binom{1}{2} = 0.$

Source Transformation

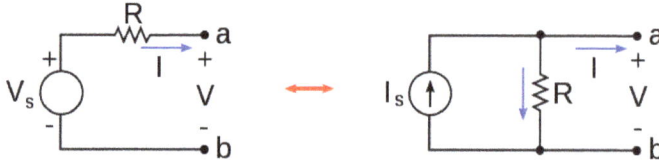

A generator with an internal impedance (i.e. non-ideal generator) can be represented as either an ideal voltage generator or an ideal current generator plus the impedance. These two forms are equivalent and the transformations are given below. If the two networks are equivalent with respect to terminals ab, then V and I must be identical for both networks. Thus,

$$V_s = RI_s \ \text{ or } I_s = \frac{V_s}{R}$$

- Norton's theorem states that any two-terminal network can be reduced to an ideal current generator and a parallel impedance.

- Thévenin's theorem states that any two-terminal network can be reduced to an ideal voltage generator plus a series impedance.

Simple Networks

Some very simple networks can be analysed without the need to apply the more systematic approaches.

Voltage Division of Series Components

Consider n impedances that are connected in series. The voltage V_i across any impedance Z_i is

$$V_i = Z_i I = \left(\frac{Z_i}{Z_1 + Z_2 + \cdots + Z_n} \right) V$$

Current Division of Parallel Components

Consider n impedances that are connected in parallel. The current I_i through any impedance Z_i is

$$I_i = \left(\frac{\left(\dfrac{1}{Z_i} \right)}{\left(\dfrac{1}{Z_1} \right) + \left(\dfrac{1}{Z_2} \right) + \cdots + \left(\dfrac{1}{Z_n} \right)} \right) I$$

for $i = 1, 2, ..., n$.

Special case: Current division of two parallel components,

$$I_1 = \left(\frac{Z_2}{Z_1 + Z_2} \right) I$$

$$I_2 = \left(\frac{Z_1}{Z_1 + Z_2} \right) I$$

Nodal Analysis

1. Label all nodes in the circuit. Arbitrarily select any node as reference.

2. Define a voltage variable from every remaining node to the reference. These voltage variables must be defined as voltage rises with respect to the reference node.

3. Write a KCL equation for every node except the reference.

4. Solve the resulting system of equations.

Mesh Analysis

Mesh — a loop that does not contain an inner loop.

1. Count the number of "window panes" in the circuit. Assign a mesh current to each window pane.

2. Write a KVL equation for every mesh whose current is unknown.

3. Solve the resulting equations.

Superposition

In this method, the effect of each generator in turn is calculated. All the generators other than the one being considered are removed and either short-circuited in the case of voltage generators or open-circuited in the case of current generators. The total current through or the total voltage across a particular branch is then calculated by summing all the individual currents or voltages.

There is an underlying assumption to this method that the total current or voltage is a linear superposition of its parts. Therefore, the method cannot be used if non-linear components are present. Note that mesh analysis and node analysis also implicitly use superposition so these too, are only applicable to linear circuits. Superposition cannot be used to find total power consumed by elements even in linear circuits. Power varies according to the square of total voltage or current and the square of the sum is not generally equal to the sum of the squares.

Choice of Method

Choice of method is to some extent a matter of taste. If the network is particularly simple or only a specific current or voltage is required then ad-hoc application of some simple equivalent circuits may yield the answer without recourse to the more systematic methods:

- Nodal analysis: The number of voltage variables, and hence simultaneous equations to solve, equals the number of nodes minus one. Every voltage source connected to the reference node reduces the number of unknowns and equations by one.

- Mesh analysis: The number of current variables, and hence simultaneous equations to solve, equals the number of meshes. Every current source in a mesh reduces the number of unknowns by one. Mesh analysis can only be used with networks which can be drawn as a planar network, that is, with no crossing components.

- Superposition is possibly the most conceptually simple method but rapidly leads to a large number of equations and messy impedance combinations as the network becomes larger.

- Effective medium approximations: For a network consisting of a high density of random resistors, an exact solution for each individual element may be impractical or impossible. Instead, the effective resistance and current distribution properties can be modelled in terms of graph measures and geometrical properties of networks.

Transfer Function

A transfer function expresses the relationship between an input and an output of a network. For resistive networks, this will always be a simple real number or an expression which boils down to a real number. Resistive networks are represented by a system of simultaneous algebraic equations. However, in the general case of linear networks, the network is represented by a system of simultaneous linear differential equations. In network analysis, rather than use the differential equations directly, it is usual practice to carry out a Laplace transform on them first and then express the result in terms of the Laplace parameter s, which in general is complex. This is described as working in the s-domain. Working with the equations directly would be described as working in the time (or t) domain because the results would be expressed as time varying quantities. The Laplace transform is the mathematical method of transforming between the s-domain and the t-domain.

This approach is standard in control theory and is useful for determining stability of a system, for instance, in an amplifier with feedback.

Two Terminal Component Transfer Functions

For two terminal components the transfer function, or more generally for non-linear elements, the constitutive equation, is the relationship between the current input to the device and the resulting voltage across it. The transfer function, Z(s), will thus have units of impedance – ohms. For the three passive components found in electrical networks, the transfer functions are;

$$\text{Resistor} \qquad Z(s) = R$$

$$\text{Inductor} \qquad Z(s) = sL$$

$$\text{Capacitor} \qquad Z(s) = \frac{1}{sC}$$

For a network to which only steady ac signals are applied, s is replaced with $j\omega$ and the more familiar values from ac network theory result.

$$\text{Resistor} \qquad Z(j\omega) = R$$

$$\text{Inductor} \qquad Z(j\omega) = j\omega L$$

$$\text{Capacitor} \qquad Z(j\omega) = \frac{1}{j\omega C}$$

Finally, for a network to which only steady dc is applied, s is replaced with zero and dc network theory applies.

$$\text{Resistor} \qquad Z = R$$

$$\text{Inductor} \qquad Z = 0$$

$$\text{Capacitor} \qquad Z = \infty$$

Two Port Network Transfer Function

Transfer functions, in general, in control theory are given the symbol H(s). Most commonly in electronics, transfer function is defined as the ratio of output voltage to input voltage and given the symbol A(s), or more commonly (because analysis is invariably done in terms of sine wave response), A(jω), so that;

$$A(j\omega) = \frac{V_o}{V_i}$$

The A standing for attenuation, or amplification, depending on context. In general, this will be a complex function of $j\omega$, which can be derived from an analysis of the impedances in the network and their individual transfer functions. Sometimes the analyst is only interested in the magnitude of the gain and not the phase angle. In this case the complex numbers can be eliminated from the transfer function and it might then be written as;

$$A(\omega) = \left| \frac{V_o}{V_i} \right|$$

Two Port Parameters

The concept of a two-port network can be useful in network analysis as a black box approach to analysis. The behaviour of the two-port network in a larger network can be entirely characterised without necessarily stating anything about the internal structure. However, to do this it is necessary to have more information than just the A(jω) described above. It can be shown that four such parameters are required to fully characterise the two-port network. These could be the forward transfer function, the input impedance, the reverse transfer function (i.e., the voltage appearing at the input when a voltage is applied to the output) and the output impedance. There are many others, one of these expresses all four parameters as impedances. It is usual to express the four parameters as a matrix;

$$\begin{bmatrix} V_1 \\ V_o \end{bmatrix} = \begin{bmatrix} z(j\omega)_{11} & z(j\omega)_{12} \\ z(j\omega)_{21} & z(j\omega)_{22} \end{bmatrix} \begin{bmatrix} I_1 \\ I_o \end{bmatrix}$$

The matrix may be abbreviated to a representative element;

$$[z(j\omega)] \text{ or just } [z]$$

These concepts are capable of being extended to networks of more than two ports. However, this is rarely done in reality because, in many practical cases, ports are considered either purely input or purely output. If reverse direction transfer functions are ignored, a multi-port network can always be decomposed into a number of two-port networks.

Distributed Components

Where a network is composed of discrete components, analysis using two-port networks is a matter of choice, not essential. The network can always alternatively be analysed in terms of its individual component transfer functions. However, if a network contains distributed components, such as in the case of a transmission line, then it is not possible to analyse in terms of individual components since they do not exist. The most common approach to this is to model the line as a two-port network and characterise it using two-port parameters (or something equivalent to them). Another example of this technique is modelling the carriers crossing the base region in a high frequency transistor. The base region has to be modelled as distributed resistance and capacitance rather than lumped components.

Analysis

Transmission lines and certain types of filter design use the image method to determine their transfer parameters. In this method, the behaviour of an infinitely long cascade connected chain of identical networks is considered. The input and output impedances and the forward and reverse transmission functions are then calculated for this infinitely long chain. Although the theoretical values so obtained can never be exactly realised in practice, in many cases they serve as a very good approximation for the behaviour of a finite chain as long as it is not too short.

Non-linear Networks

Most electronic designs are, in reality, non-linear. There is very little that does not include some semiconductor devices. These are invariably non-linear, the transfer function of an ideal semiconductor p-n junction is given by the very non-linear relationship;

where,

- i and v are the instantaneous current and voltage.

- I_o is an arbitrary parameter called the reverse leakage current whose value depends on the construction of the device.

- V_T is a parameter proportional to temperature called the thermal voltage and equal to about 25mV at room temperature.

There are many other ways that non-linearity can appear in a network. All methods utilising linear superposition will fail when non-linear components are present. There are several options for dealing with non-linearity depending on the type of circuit and the information the analyst wishes to obtain.

Constitutive Equations

The diode equation above is an example of an element constitutive equation of the general form,

$$f(v,i) = 0$$

This can be thought of as a non-linear resistor. The corresponding constitutive equations for non-linear inductors and capacitors are respectively;

$$f(v,\varphi)=0$$
$$f(v,q)=0$$

where, f is any arbitrary function, φ is the stored magnetic flux and q is the stored charge.

Existence, Uniqueness and Stability

An important consideration in non-linear analysis is the question of uniqueness. For a network composed of linear components there will always be one, and only one, unique solution for a given set of boundary conditions. This is not always the case in non-linear circuits. For instance, a linear resistor with a fixed current applied to it has only one solution for the voltage across it. On the other hand, the non-linear tunnel diode has up to three solutions for the voltage for a given current. That is, a particular solution for the current through the diode is not unique, there may be others, equally valid. In some cases there may not be a solution at all: the question of existence of solutions must be considered.

Another important consideration is the question of stability. A particular solution may exist, but it may not be stable, rapidly departing from that point at the slightest stimulation. It can be shown that a network that is absolutely stable for all conditions must have one, and only one, solution for each set of conditions.

Methods

Boolean Analysis of Switching Networks

A switching device is one where the non-linearity is utilised to produce two opposite states. CMOS devices in digital circuits, for instance, have their output connected to either the positive or the negative supply rail and are never found at anything in between except during a transient period when the device is actually switching. Here the non-linearity is designed to be extreme, and the analyst can actually take advantage of that fact. These kinds of networks can be analysed using Boolean algebra by assigning the two states ("on"/"off", "positive"/"negative" or whatever states are being used) to the boolean constants "0" and "1".

The transients are ignored in this analysis, along with any slight discrepancy between the actual state of the device and the nominal state assigned to a boolean value. For instance, boolean "1" may be assigned to the state of +5V. The output of the device may actually be +4.5V but the analyst still considers this to be boolean "1". Device manufacturers will usually specify a range of values in their data sheets that are to be considered undefined (i.e. the result will be unpredictable).

The transients are not entirely uninteresting to the analyst. The maximum rate of switching is determined by the speed of transition from one state to the other. Happily for the analyst, for many devices most of the transition occurs in the linear portion of the devices transfer function and linear analysis can be applied to obtain at least an approximate answer.

It is mathematically possible to derive boolean algebras which have more than two states. There is not too much use found for these in electronics, although three-state devices are passingly common.

Separation of Bias and Signal Analyses

This technique is used where the operation of the circuit is to be essentiallylinear, but the devices used to implement it are non-linear. A transistor amplifier is an example of this kind of network. The essence of this technique is to separate the analysis into two parts. Firstly, the dc biases are analysed using some non-linear method. This establishes the quiescentoperating point of the circuit. Secondly, the small signal characteristics of the circuit are analysed using linear network analysis.

Graphical Method of DC Analysis

In a great many circuit designs, the DC bias is fed to a non-linear component via a resistor (or possibly a network of resistors). Since resistors are linear components, it is particularly easy to determine the quiescent operating point of the non-linear device from a graph of its transfer function. The method is as follows: from linear network analysis the output transfer function (that is output voltage against output current) is calculated for the network of resistor(s) and the generator driving them. This will be a straight line (called the load line) and can readily be superimposed on the transfer function plot of the non-linear device. The point where the lines cross is the quiescent operating point.

Perhaps the easiest practical method is to calculate the (linear) network open circuit voltage and short circuit current and plot these on the transfer function of the non-linear device. The straight line joining these two point is the transfer function of the network.

In reality, the designer of the circuit would proceed in the reverse direction to that described. Starting from a plot provided in the manufacturers data sheet for the non-linear device, the designer would choose the desired operating point and then calculate the linear component values required to achieve it.

It is still possible to use this method if the device being biased has its bias fed through another device which is itself non-linear – a diode for instance. In this case however, the plot of the network transfer function onto the device being biased would no longer be a straight line and is consequently more tedious to do.

Small Signal Equivalent Circuit

This method can be used where the deviation of the input and output signals in a network stay within a substantially linear portion of the non-linear devices transfer function, or else are so small that the curve of the transfer function can be considered linear. Under a set of these specific conditions, the non-linear device can be represented by an equivalent linear network. It must be remembered that this equivalent circuit is entirely notional and only valid for the small signal deviations. It is entirely inapplicable to the dc biasing of the device.

For a simple two-terminal device, the small signal equivalent circuit may be no more than two components. A resistance equal to the slope of the v/i curve at the operating point (called the dynamic resistance), and tangent to the curve. A generator, because this tangent will not, in general, pass through the origin. With more terminals, more complicated equivalent circuits are required.

A popular form of specifying the small signal equivalent circuit amongst transistor manufacturers is to use the two-port network parameters known as (h) parameters. These are a matrix of four parameters as with the (z) parameters but in the case of the (h) parameters they are a hybrid mixture of impedances, admittances, current gains and voltage gains. In this model the three terminal transistor is considered to be a two port network, one of its terminals being common to both ports. The (h) parameters are quite different depending on which terminal is chosen as the common one. The most important parameter for transistors is usually the forward current gain, h_{21}, in the common emitter configuration. This is designated h_{fe} on data sheets.

The small signal equivalent circuit in terms of two-port parameters leads to the concept of dependent generators. That is, the value of a voltage or current generator depends linearly on a voltage or current elsewhere in the circuit. For instance the (z) parameter model leads to dependent voltage generators as shown in this diagram;

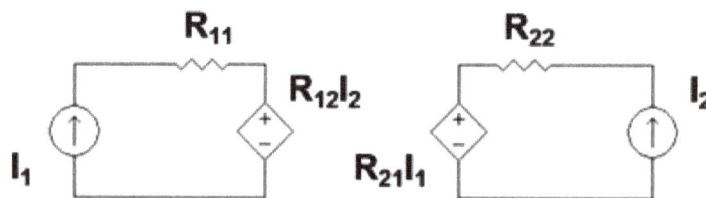

(z) Parameter equivalent circuit showing dependent voltage generators

There will always be dependent generators in a two-port parameter equivalent circuit. This applies to the (h) parameters as well as to the (z) and any other kind. These dependencies must be preserved when developing the equations in a larger linear network analysis.

Piecewise Linear Method

In this method, the transfer function of the non-linear device is broken up into regions. Each of these regions is approximated by a straight line. Thus, the transfer function will be linear up to a particular point where there will be a discontinuity. Past this point the transfer function will again be linear but with a different slope.

A well known application of this method is the approximation of the transfer function of a pn junction diode. The actual transfer function of an ideal diode has been given at the top of this (non-linear) section. However, this formula is rarely used in network analysis, a piecewise approximation being used instead. It can be seen that the diode current rapidly diminishes to $-I_0$ as the voltage falls. This current, for most purposes, is so small it can be ignored. With increasing voltage, the current increases exponentially. The diode is modelled as an open circuit up to the knee of the exponential curve, then past this point as a resistor equal to the bulk resistance of the semiconducting material.

The commonly accepted values for the transition point voltage are 0.7V for silicon devices and 0.3V for germanium devices. An even simpler model of the diode, sometimes used in switching applications, is short circuit for forward voltages and open circuit for reverse voltages.

The model of a forward biased pn junction having an approximately constant 0.7V is also a much used approximation for transistor base-emitter junction voltage in amplifier design.

The piecewise method is similar to the small signal method in that linear network analysis techniques can only be applied if the signal stays within certain bounds. If the signal crosses a discontinuity point then the model is no longer valid for linear analysis purposes. The model does have the advantage over small signal however, in that it is equally applicable to signal and dc bias. These can therefore both be analysed in the same operations and will be linearly superimposable.

Time-varying Components

In linear analysis, the components of the network are assumed to be unchanging, but in some circuits this does not apply, such as sweep oscillators, voltage controlled amplifiers, and variable equalisers. In many circumstances the change in component value is periodic. A non-linear component excited with a periodic signal, for instance, can be represented as a periodically varying *linear* component. Sidney Darlington disclosed a method of analysing such periodic time varying circuits. He developed canonical circuit forms which are analogous to the canonical forms of Ronald M. Foster and Wilhelm Cauer used for analysing linear circuits.

Vector Circuit Theory

Generalization of circuit theory based on scalar quantities to vectorial currents is a necessity for newly evolving circuits such as spin circuits. Generalized circuit variables consist of four components: scalar current and vector spin current in x, y, and z directions. The voltages and currents each become vector quantities with conductance described as a 4x4 spin conductance matrix.

Mesh Analysis

Mesh is a loop that doesn't consists of any other loop inside it. Mesh analysis technique, uses mesh currents as variables, instead of currents in the elements to analyse the circuit. Therefore, this method absolutely reduces the number of equations to be solved. Mesh analysis applies the Kirchhoff's Voltage Law (KVL) to determine the unknown currents in a given circuit. Mesh analysis is also called as mesh-current method or loop analysis. After finding the mesh currents using KVL, voltages anywhere in a given circuit can be determined by using Ohms law.

Steps to Analyse the Mesh Analysis Technique

1) Check whether there is a possibility to transform all current sources in the given circuit to voltage sources.

2) Assign the current directions to each mesh in a given circuit and follow the same direction for each mesh.

3) Apply KVL to each mesh and simplify the KVL equations.

4) Solve the simultaneous equations of various meshes to get the mesh currents and these equations are exactly equal to the number of meshes present in the network.

Consider the below DC circuit to apply the mesh current analysis, such that currents in different meshes can be found. In the below figure there are three meshes present as ACDA, CBDC and ABCA but the path ABDA is not a mesh. As a first step, the current through each mesh is assigned with the same direction as shown in figure,

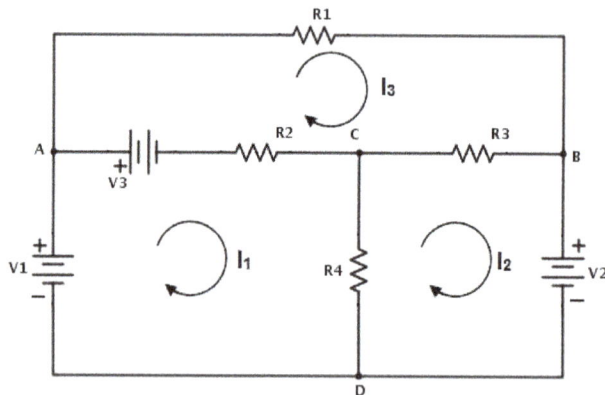

Secondly, for each mesh we have to apply KVL. By applying KVL around the first loop or mesh we get

$$V_1 - V_3 - R_2 (I_1 - I_3) - R_4 (I_1 - I_2) = 0$$

$$V_1 - V_3 = I_1 (R_2 + R_4) - I_2 R_4 - I_3 R_2$$

Similarly, by applying KVL around second mesh we get,

$$-V_2 - R_3(I_2 - I_3) - R_4(I_2 - I_1) = 0$$

$$-V_2 = -I_1R_4 + I_2(R_3 + R_4) - I_3R_3$$

And by applying KVL around third mesh or loop we get,

$$V_3 - R_1I_3 - R_3(I_3 - I_2) - R_2(I_3 - I_1) = 0$$

$$V_3 = -I_1R_2 - I_2R_3 + I_3(R_1 + R_2 + R_3)$$

Therefore, by solving the above three equations we can obtain the mesh currents for each mesh in the given circuit.

Example problems on mesh analysis:

Example:

Consider the below example in which we find the voltage across the 12A current source using mesh analysis. In the given circuit all the sources are current sources.

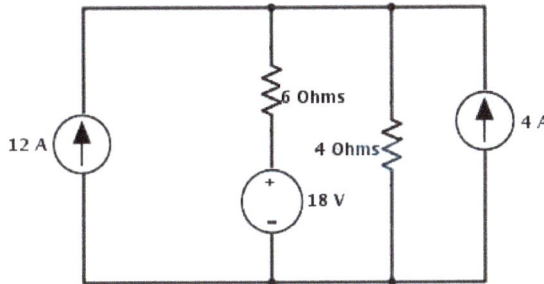

Step 1: In the circuit there is a possibility to change the current source to a voltage source on right hand side source with parallel resistance. The current source is converted into a voltage source by placing the same value of resistor in series with a voltage source and the voltage in that source is determined as,

$$V_s = I_s \times R_s$$

$$= 4 \times 4 = 16 \text{ Volts}$$

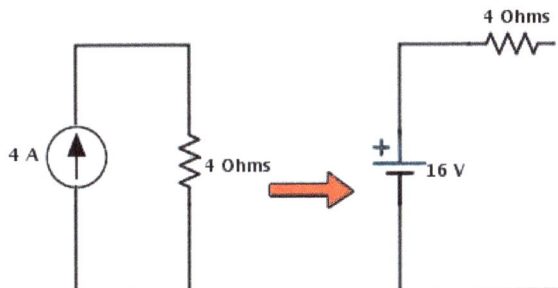

Step 2: Assign the branch currents as I_1 and I_2 to the respective branches or loops and represent the direction of currents as shown below:

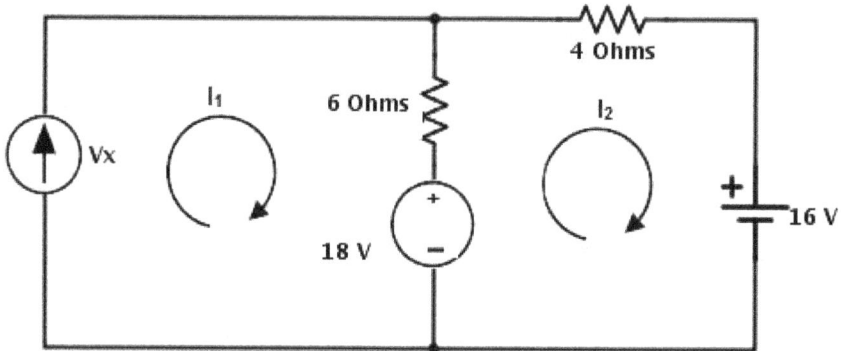

Step 3: Apply the KVL to each mesh in the given circuit

Mesh -1:

$$V_x - 6 \times (I_1 - I_2) - 18 = 0$$

Substituting I1 = 12 A

$$V_x + 6_{I2} = 90 .... (1)$$

Mesh – 2:

$$18 - 6 \times (I_2 - I_1) - 4 \times I_2 - 16 = 0$$

$$2 - 10 \times I_2 + 6(12) = 0$$

I2 = 74/ 10

= 7.4 Amps

Substituting in equation 1 we get

$$V_x = 90 - 44.4$$

= 45.6 Volts

Example:

Consider the below circuit where we determine the voltage across the current source and a branch current Iac. Assign the directions as shown below and note that current is assigned opposite to the source current in second loop.

By applying KVL to the first mesh we get

$$V_1 - R_2 (I_1 - I_3) - R_4 (I_1 - I_2) = 0$$

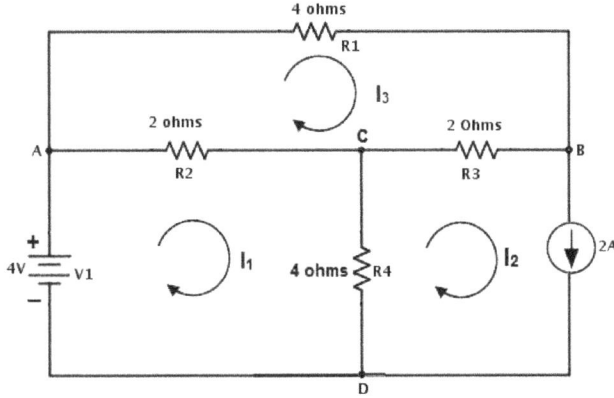

$$4 - 2I_1 - 2I_3 - 4I_1 - 4I_2 = 0$$

$$-6I_1 - 2I_3 = 4$$

By applying KVL to the second mesh we get

$$-V_c - R_4(I_2 - I_1) - R_3(I_2 - I_3) = 0$$

$$-V_c = 4I_2 - 4I_1 + 2I_2 - 2I_3 = 0$$

$$-V_c = -4I_1 + 6I_2 - 2I_3$$

But I_2 = -2 A, then

$$-V_c = -4I_1 - 12 - 2I_3$$

By applying KVL to the third mesh we get

$$-R_1I_3 - R_3(I_3 - I_2) - R_2(I_3 - I_1) = 0$$

$$-4I_3 - 2I_3 + 2I_2 - 2I_3 + 2I_1 = 0$$

$$-8I_3 - 4 + 2I_1 = 0 \text{ (by substituting } I_2 = \text{-2 A)}$$

$$2I_1 - 8I_3 = 4$$

By solving 1 and 3 equations we get I_3 = -0.615 and I_1 = 4.46

Therefore, the voltage $V_c = 4(4.46) + 12 + 2(-0.615)$

$$V_c = 28.61 \text{ V}$$

And the branch current $I_{ac} = I_1 - I_3$

$$I_{ac} = 5.075 \text{ amps}$$

Likewise we can find every branch current using the mesh analysis.

Super Mesh Transform

A super mesh is formed when two adjacent meshes share a common current source and none of these (adjacent) meshes contains a current source in the outer loop. Consider the below circuit in which super mesh is formed by the loop around the current source,

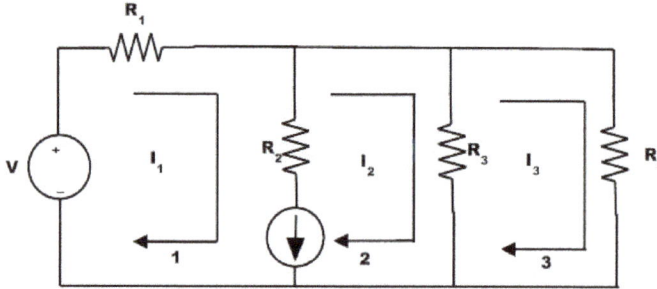

The current source is common to the meshes 1 and 2 and hence it must be analysed independently. To achieve this, assume the branch that contains current source is open circuited and create a new mesh called super mesh.

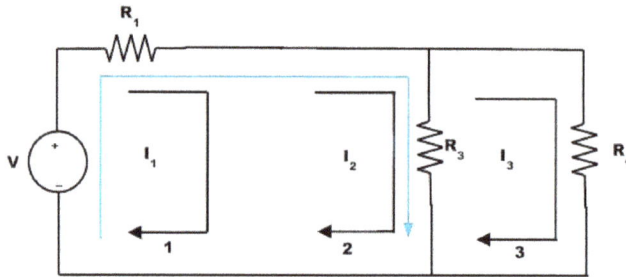

Writing KVL to the super mesh we get,

$$V = I_1 R_1 + (I_2 - I_3) R_3$$

$$= I_1 R_1 + I_2 R_3 - I_3 R_3$$

Applying KVL to the Mesh 3 we get,

$$(I_3 - I_2) R_3 + I_3 R_4 = 0$$

And the difference between the two mesh currents gives the current from the current source. Here the current source direction is in the loop current direction I_1. Hence I_1 is more than I_2, then,

$$I = I_1 - I_2$$

Thus, by using these three mesh equations we can easily find the three unknown currents in the network.

Example on Supermesh Analysis

Consider the below example in which we have to find the current through the 10 ohms resistor,

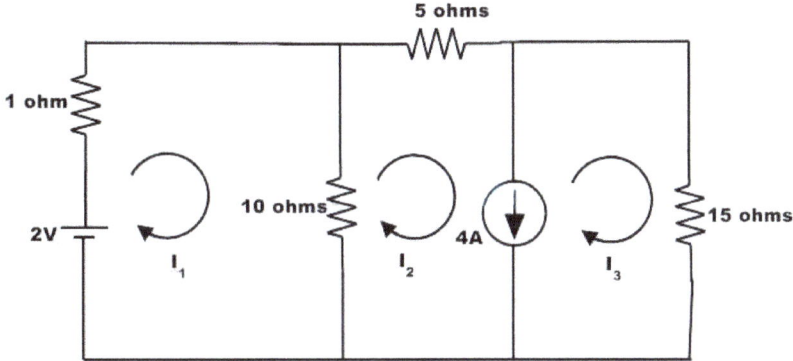

By applying the KVL to the mesh 1 we get,

$$1I_1 + 10\,(I_1 - I_2) = 2$$

$$11I_1 - 10\,I_2 = 2$$

The meshes 2 and 3 consist of 4A current source and hence form a super mesh. The current from 4A current source is in the direction of I_3 and thus the super mesh current is given as,

$$I = I_3 - I_2$$

$$I_3 - I_2 = 4$$

By applying KVL to the outer loop of the super mesh we get,

$$-10\,(I_2 - I_1) - 5I_2 - 15I_3 = 0$$

$$10I_1 - 15I_2 - 15I_3 = 0$$

By solving 1, 2 and 3 equations, we get

$$I_1 = -2.35 \text{ A}$$

$$I_2 = -2.78 \text{ A}$$

$$I_3 = 1.22 \text{ A}$$

Hence the current through the 10 ohms resistor is $I_1 - I_2$,

$$= -2.35 + 2.78 \text{ A}$$

$$= 0.43 \text{ A}$$

Nodal Analysis

The main concept behind the nodal analysis is that, in a given circuit if the node voltages are known, then we can immediately determine all branch currents associated with the circuit. As we know that, for finding node voltages we use KCL. In this technique, node voltages are considered as variables in the circuit, instead of element voltages, which results in reduction of the number of equations to simplify the circuit. In nodal analysis method, with availability of all the nodes, one node is considered as a reference node (zero potential) and it is represented as ground terminal. To other remaining unknown nodes, voltages are assigned, with respect to the referenced node voltage.

To the each node in a given circuit, we apply the KCL except for the referenced node. Suppose if the given circuit has N nodes, then we get N-1 simultaneous equations to find the N-1 unknown node voltages.

Steps to Analyse Nodal Analysis Technique

1) Check the possibility to transform voltage sources in the given circuit to the current sources and transform them.

2) Identify the nodes present in the given circuit and assign one node as reference node and with respect to this ground or reference node, label other nodes as unknown node voltages.

3) Assign the current direction in each branch in the given circuit (it is an arbitrary decision).

4) Apply KCL to N-1 nodes and write nodal equations by expressing the branch currents as node assigned voltages.

5) Solve the simultaneous equations of nodes to find the node voltages and finally branch currents. The number of node equations is equal to the number of nodes minus one (as one node is referenced).

Consider below DC circuit, in which branch currents are to be determined using the nodal analysis.

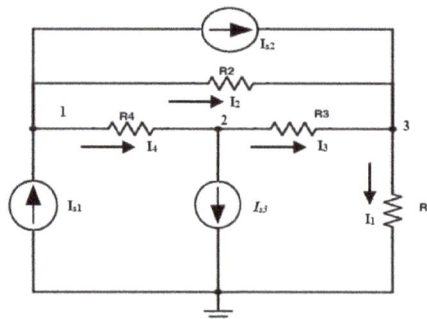

As a first step in nodal analysis, we have to select the reference node which is to be connected to the zero or ground potential as indicated below.

Secondly, we apply Kirchoff's Current Law (KCL) to each node in the circuit except the reference node. By applying KCL at node 1 we get,

$$I_{s1} - I_{s3} - I_4 - I_2 = 0$$

$$I_{s1} - I_{s3} - \{(V_1 - V_2)/R_4\} - \{(V_1 - V_3)/R_2\} = 0$$

$$I_{s1} - I_{s3} = V_1 \{ (1/R_2) + (1/R_4)\} - V_2 (1/R_4) - V_3 (1/R_2)$$

$$I_{s1} - I_{s3} = G_{11} V_1 - G_{12} V_2 - G_{13} V_3$$

where, G_{ii} is the sum of total conductance at the first node. (As we know that $1/R = G$)

By applying KCL at node 2 we get

$$I_4 - I_{s2} - I_3 = 0$$

$$\{(V_1 - V_2)/R_4\} - I_{s2} - \{(V_2 - V_3)/R_3\} = 0$$

$$- I_{s2} = - V_1 (1/R_4) + V_2 \{ (1/R_3) + (1/R_4)\} - V_3 (1/R_3)$$

$$- I_{s2} = - G_{21} V_1 - G_{22} V_2 - G_{23} V_3$$

Applying KCL at node 3 we get

$$I_{s3} + I_2 + I_3 - I_1 = 0$$

$$I_{s3} + \{(V_1 - V_3)/R_2\} - \{(V_2 - V_3)/R_3\} - V_3 (1/R_1) = 0$$

$$I_{s3} = - V_1 (1/R_2) - V_2 (1/R_3) + V_3 \{ (1/R_1) + (1/R_2) + (1/R_3)\}$$

$$I_{s3} = - G_{31} V_1 - G_{32} V_2 + G_{33} V_3$$

Likewise, we can write the KCL equations for i th node. And hence

$\sum I_{ii}$ is equal to the algebraic sum of all the currents connected at the i th node where i = 1, 2, 3......N and N = n-1 (n is the total number of nodes present in the circuit).

G_{ii} = The sum of conductance connected to the i th node.

G_{ij} = The sum of conductance connected between i and j nodes.

By solving the above three equations we get the branch voltages at respective nodes and thereby we can calculate branch currents.

Example

Determine the node voltages and currents in each branch using nodal analysis method in the given circuit.

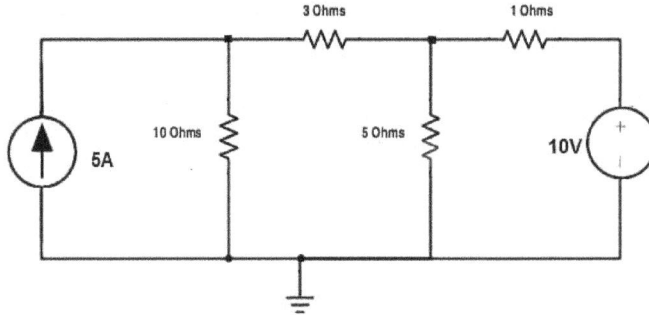

The given circuit contains a voltage source. This can be transformed to current source or can be analysed directly without any transformation. Now let us calculate the nodal voltages without any transformation.

As a first step in nodal analysis, we have to choose and label the nodes present in the given circuit. By choosing the bottom node as reference node, we have two another nodes in the given circuit. So these nodes are labelled as V1 and V2 as shown in below figure. And also current directions in each branch are represented.

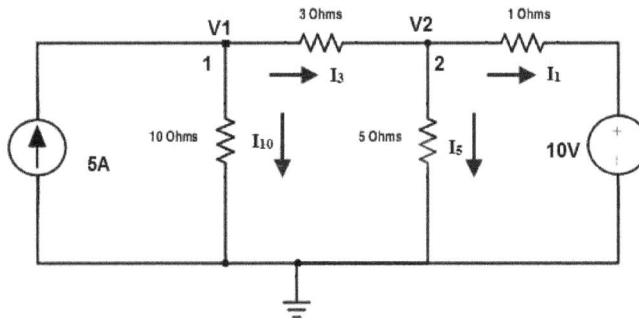

By applying KCL at node 1, we get

$$5 = I_3 + I_{10}$$

$$5 = (V_1/10) + (V_1 - V_{2/3})$$

$$13V_1 - 10V_2 = 150$$

By applying KCL at node 2, we get

$$I_3 = I_5 + I_1$$

$$(V_1 - V_{2/3}) = (V_{2/5}) + (V_2 - V_{10/1})$$

$$5V_1 - 23V_2 = -150$$

By solving above two equations, we get

$$V_1 = 19.85 \text{ Volts and } V_2 = 10.9 \text{ Volts}$$

The currents in each branch is given as

$$I_{10} = V_{1/10}$$

$$= 19.85/10 = 1.985$$

$$I_3 = V_1 - V_{2/3}$$

$$= 19.85 - 10.9/3$$

$$= 2.98 \text{ A}$$

$$I_5 = V_2/5$$

$$= 10.9/5$$

$$= 2.18 \text{ A}$$

$$I_1 = V_2 - 10$$

$$= 10.9 - 10$$

$$= 0.9 \text{ A}$$

Distributed Element Model

In electrical engineering, the distributed element model or transmission line model of electrical circuits assumes that the attributes of the circuit (resistance, capacitance, and inductance) are distributed continuously throughout the material of the circuit. This is in contrast to the more common lumped element model, which assumes that these values are lumped into electrical components that are joined by perfectly conducting wires. In the distributed element model, each circuit element is infinitesimally small, and the wires connecting elements are not assumed to be perfect conductors; that is, they have impedance. Unlike the lumped element model, it assumes non-uniform current along each branch and non-uniform voltage along each node. The distributed model is used at high frequencies.

Bartlett's Bisection Theorem

Bartlett's bisection theorem states that if a symmetrical network is bisected and one half is impedance-scaled including the termination, the response shape will not change. All odd-order Butterworth and odd-order Tchebyscheff filters satisfy this requirement, and also 3rd order inverse Tchebyscheff and 3rd order elliptic filters. The symmetry must include not only the topology but also the component values, therefore higher order inverse Tchebyscheff and elliptic filters are excluded. This extremely powerful result allows many passive filters to be modified for unequal source and load resistances.

Example: Modify the circuit in figure for a termination resistance of 1000 ohms.

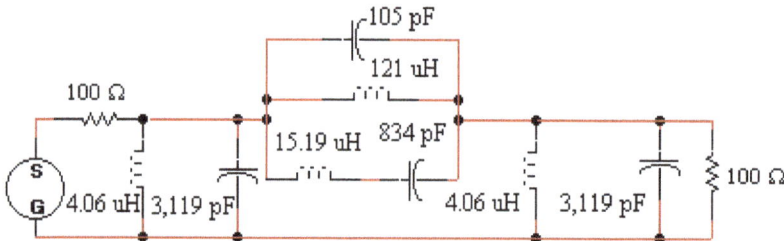

Figure: Bandpass 3rd order elliptic filter with 1dB passband ripple, 50dB stopband attenuation and a bandpass region of 1MHz to 2MHz.and load resistances.

In order to make the component values symetrical, the 105 pF capacitor is split into two series 210 pF capacitors. The 121 uH inductor is split into two series 60.5 uH inductors. The 15.19 uH inductor in series with the 834 pF capacitor is split into two 7.6 uH series inductors and two 1668 pF series capacitors. Those components on the right side of the dividing line in figure are impedance scaled to increase their impedance by a factor of 10. The capacitor values therefore decrease by a factor of 10, and the inductor values are increased by a factor of 10.

Solution: The bisected circuit becomes

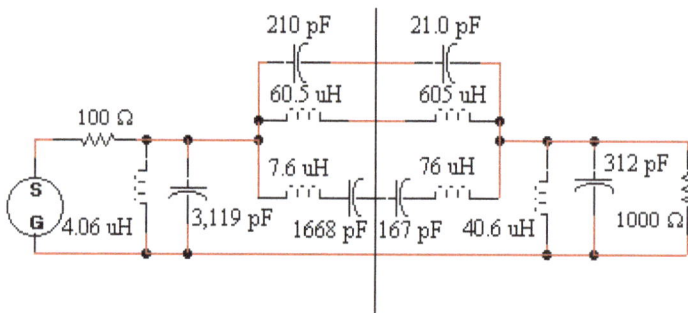

Figure: Bisected circuit

Combining the series inductors and capacitors yields

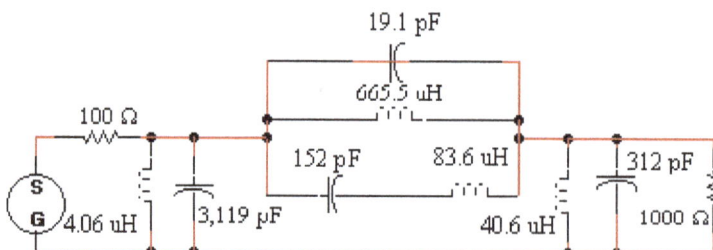

Figure: Final circuit after modification for a 100 ohm source impedance and 1000 ohm load impedance.

Bartlett's bisection theorem cannot be used to provide an impedance match between different source and load resistances but the transformed network will have the exact same response shape and phase as the original network.

Delta-wye Transform and Wye-delta Transform

The Delta-Wye transformation is an extra technique for transforming certain resistor combinations that cannot be handled by the series and parallel equations. This is also referred to as a P_i - T transformation.

Sometimes when you are simplifying a resistor network, you get stuck. Some resistor networks cannot be simplified using the usual series and parallel combinations. This situation can often be handled by trying the $\Delta-Y$ transformation, or 'Delta-Wye' transformation.

The names Delta and Wye come from the shape of the schematics, which resemble letters. The transformation allows you to replace three resistors in a Δ configuration by three resistors in a Y configuration, and the other way around.

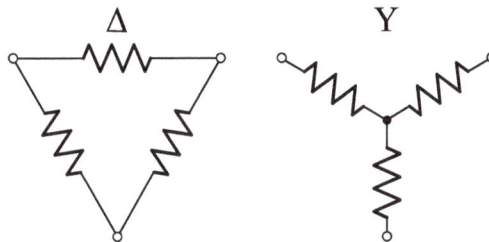

The $\Delta-Y$ drawing style emphasizes these are 3-terminal configurations. Something to notice is the different number of nodes in the two configurations. Δ has three nodes, while Y has four nodes (one extra in the center).

The configurations can be redrawn to square up the resistors. This is called a $\pi-T$ configuration,

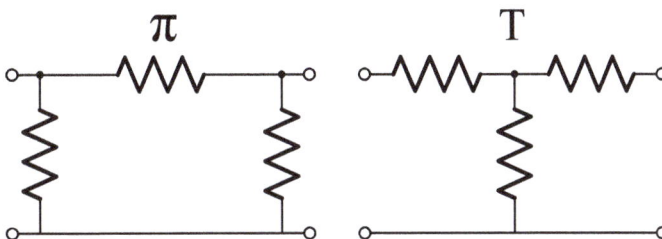

The $\pi-T$ style is a more conventional drawing you would find in a typical schematic. The transformation equations developed next apply to $\pi-T$ as well.

Δ–Ydelta, minus, Y transformation

For the transformation to be equivalent, the resistance between each pair of terminals must be the same before and after. It is possible to write three simultaneous equations to capture this constraint.

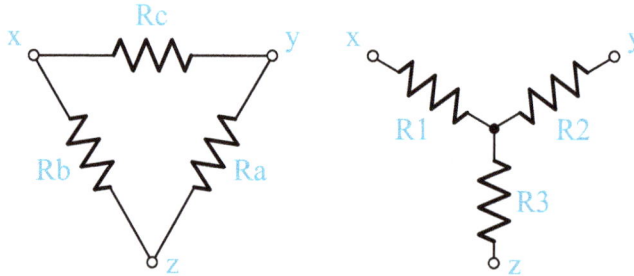

Consider terminals x and y (and for the moment assume terminal z isn't connected to anything, so the current in R_3 is 0). In the Δ configuration, the resistance between x and y is Rc, in parallel with Ra +Rb.

On the Y side, the resistance between x and y is the series combination $R_1 + R_2$ (again, assume terminal z isn't connected to anything, so R_1 and R_2 carry the same current and can be considered in series). We set these equal to each other to get the first of three simultaneous equations,

$$R_1 + R_2 = \frac{Rc(Ra + Rb)}{Rc + (Ra + Rb)}$$

We can write two similar expressions for the other two pairs of terminals. Notice the Δ resistors have letter names, (Ra, etc.) Y resistors have number names, (R_1, etc.).

After solving the simultaneous equations (not shown), we get the equations to transform either network into the other.

Δ→Y transformation

Equations for transforming a Δ network into a Y network:

$$R_1 = \frac{Rb\,Rc}{Ra + Rb + Rc}$$

$$R_2 = \frac{Ra\,Rc}{Ra + Rb + Rc}$$

$$R_3 = \frac{Ra\,Rb}{Ra + Rb + Rc}$$

Transforming from Δ to Y introduces one additional node.

Y→Δ transformation

Equations for transforming a Y network into a Δ network:

$$Ra = \frac{R_1 R_2 + R_2 R_3 + R_3 R_1}{R_1}$$

$$Rb = \frac{R_1 R_2 + R_2 R_3 + R_3 R_1}{R_2}$$

$$Rc = \frac{R_1 R_2 + R_2 R_3 + R_3 R_1}{R_3}$$

Transforming from Y to Δ removes one node.

Example

Let's do a symmetric example. Assume we have a Δ circuit with 3Ω resistors. Derive the Y equivalent by using the Δ→Y equations.

$$R_1 = \frac{Rb\,Rc}{Ra + Rb + Rc} = \frac{3 \cdot 3}{3 + 3 + 3} = 1\Omega$$

$$R_2 = \frac{Ra\,Rc}{Ra + Rb + Rc} = \frac{3 \cdot 3}{3 + 3 + 3} = 1\Omega$$

$$R_3 = \frac{Ra\,Rb}{Ra + Rb + Rc} = \frac{3 \cdot 3}{3 + 3 + 3} = 1\Omega$$

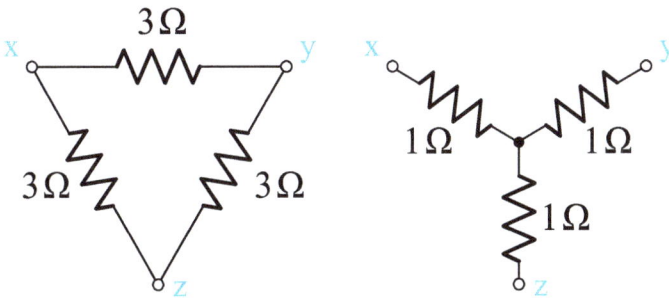

Going in the other direction, from Y→Δ, looks like this,

$$Ra = \frac{R_1 R_2 + R_2 R_3 + R_3 R_1}{R_1} = \frac{1.1 + 1.1 + 1.1}{1} = 3\Omega$$

$$Rb = \frac{R_1 R_2 + R_2 R_3 + R_3 R_1}{R_2} = \frac{1.1 + 1.1 + 1.1}{1} = 3\Omega$$

$$Rc = \frac{R_1 R_2 + R_2 R_3 + R_3 R_1}{R_3} = \frac{1.1 + 1.1 + 1.1}{1} = 3\Omega$$

Now for an example that's a little less tidy. We want to find the equivalent resistance between the top and bottom terminals.

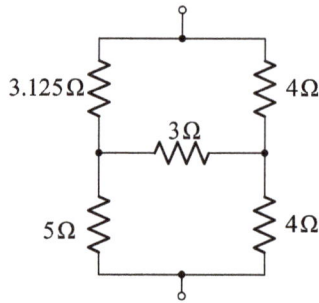

Try as we might, there are no resistors in series or in parallel. But we are not stuck. First, let's redraw the schematic to emphasize we have two Δ connections stacked one on the other.

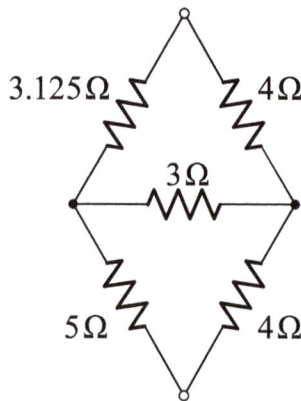

Now select one of the Δ's to convert to a Y. We will perform a Δ→Y, right arrow, Y transformation and see if it breaks the logjam, opening up other opportunities for simplification.

We go to work on the bottom Δ (an arbitrary choice). Very carefully label the resistors and nodes. To get the right answers from the transformation equations, it is critical to keep the resistor names and node names straight. Rc, c must connect between nodes x and y, and so on for the other resistors. Refer to Diagram 1 above for the labeling convention.

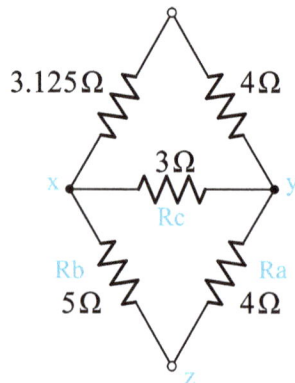

When we perform the transform on the lower Δ, the black Δ resistors will be replaced by the new gray Y resistors, like this:

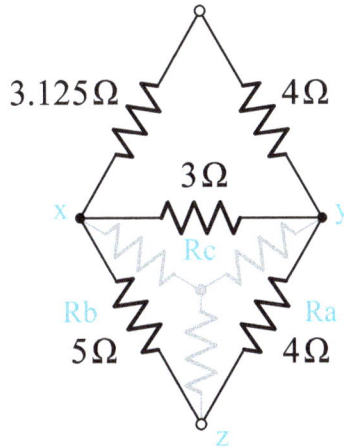

Perform the transform yourself before looking at the answer. Check that you select the right set of equations.

Compute three new resistor values to convert the Δ to a Y, and draw the complete circuit.

The circuit now has series and parallel resistors where it had none before. Continue simplification with series and parallel combinations until we get down to a single resistor between the terminals. Redraw the schematic again to square up the symbols into a familiar style.

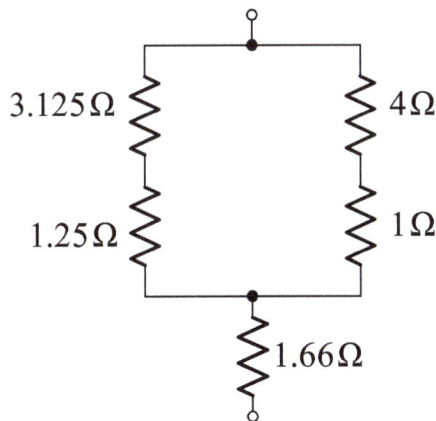

We proceed through the remaining simplification steps just as we did before in the topic on Resistor Network Simplification.

On the left branch, $3.125 + 1.25 = 4.375 \ \Omega$

On the right branch, $4 + 1 = 5\Omega$

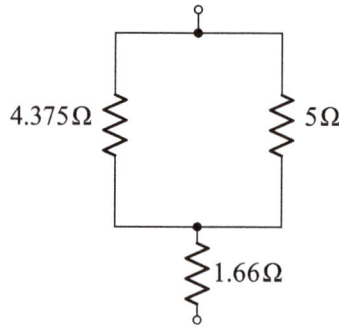

The two parallel resistors combine as $4.375 \,\|\, 5 = \dfrac{4.375 \cdot 5}{4.375 + 5} = 2.33\Omega$

And we finish by adding the last two series resistors together,

Requivalent=2.33+1.66=4Ω

Symbolic Circuit Analysis

Symbolic circuit analysis is a formal technique of circuit analysis to calculate the behaviour or characteristic of an electric/electronic circuit with the independent variables (time or frequency), the dependent variables (voltages and currents), and (some or all of) the circuit elements represented by symbols. Symbolic circuit analysis is concerned with obtaining those relationships in symbolic form, i.e., in the form of analytical expression, where the complex frequency (or time) and some or all of the circuit components are represented by symbols.

Network Analyzer

Electrical networks can be measured and assessed using the electrical network analyser. Of special interest when using this network analyser are the electrical networks' properties that pertain to reflection of electrical signals and subsequent transmission - these are called the S-parameters or scattering parameters. The usual operating frequencies of this network analyser start at nine kilohertz and reach up to 110 gigahertz.

There are however some types of network analysers which can operate at a low-frequency range reaching down to 10 hertz. A network analyser that falls in this category can help make open loop stability analysis or gauge the capacity of ultra sonic and audio parts.

A network analyser can either be a Scalar Network Analyser (or SNA) that is only devoted to gauging amplitude properties, or a Vector Network Analyser or (VNA) that encompasses amplitude plus phase properties at the same time. The VNA is sometimes dubbed an Automatic Network Analyzer or gain-phase meter. On the other hand, the SNA has the same functions as a spectrum analyzer, fused with the tracking generator. But a network analyser will probably be a VNA - so if you are not sure what type of network analyser you have, it is probably a VNA.

You might also run across a new variety of network analyser which is called the Microwave Transition Analyzer (or MTA) - sometimes known as the Large Signal Network Analyser (or LSNA) - that can gauge the phase and amplitude of the harmonics and fundamental at the same time.

The AC Network Analyser served as the model for studying major alternating current power networks between years 1929 to the latter half of the 1960s. The AC Network Analyser was actually based on the DC calculating board common to pioneering analysis of power systems. Though the AC Network Analyser were applied greatly to consideration of system stability, short circuit assessments, and analysis of power flow, they were phased out eventually in favor of numerical solutions computed by digital computers.

Continuity Test

Continuity testing is the act of testing the resistance between two points. If there is very low resistance (less than a few Ωs), the two points are connected electrically, and a tone is emitted. If there is more than a few Ωs of resistance, than the circuit is open, and no tone is emitted. This test helps insure that connections are made correctly between two points. This test also helps us detect if two points are connected that should not be.

Continuity is quite possibly the single most important function for embedded hardware gurus. This feature allows us to test for conductivity of materials and to trace where electrical connections have been made or not made.

Set the multimeter to 'Continuity' mode. It may vary among DMMs, but look for a diode symbol with propagation waves around it (like sound coming from a speaker).

Multimeter is set to continuity mode

Now touch the probes together. The multimeter should emit a tone (Note: Not all multimeters have a continuity setting, but most should). This shows that a very small amount of current is allowed to flow without resistance (or at least a very very small resistance) between probes.

On a breadboard that is not powered, use the probes to poke at two separate ground pins. You should hear a tone indicating that they are connected. Poke the probes from the VCC pin on a microcontroller to VCC on your power supply. It should emit a tone indicating that power is free to flow from the VCC pin to the micro. If it does not emit a tone, then you can begin to follow the route that copper trace takes and tell if there are breaks in the line, wire, breadboard, or PCB.

Continuity is a great way to test if two SMD pins are touching. If your eyes can't see it, the multimeter is usually a great second testing resource.

When a system is not working, continuity is one more thing to help troubleshoot the system. Here are the steps to take:

1. If the system is on, carefully check VCC and GND with the voltage setting to make sure the voltage is the correct level. If the 5V system is running at 4.2V check your regulator carefully, it could be very hot indicating the system is pulling too much current.

2. Power the system down and check continuity between VCC and GND. If there is continuity (if you hear a beep), then you've got a short somewhere.

3. Power the system down. With continuity, check that VCC and GND are correctly wired to the pins on the microcontroller and other devices. The system may be powering up, but the individual ICs may be wired wrong.

4. Assuming you can get the microcontroller running, set the multimeter aside, and move on to serial debugging or use a logic analyzer to inspect the digital signals.

Continuity and large capacitors: During normal troubleshooting. you will be probing for continuity between ground and the VCC rail. This is a good sanity check before powering up a prototype to make sure there is not a short on the power system. But don't be surprised if you hear a short 'beep' when probing. This is because there is often significant amounts of capacitance on the power system. The multimeter is looking for very low resistance to see if two points are connected. Capacitors will act like a short for a split second until they fill up with energy, and then act like an open connection. Therefore, you will hear a short beep and then nothing.

Large-signal Model

A large signal model of a given circuit is based on the assumption that the large signals actually affect the operating point or biasing of the given circuit, in this model the elements are not linearized i.e. Elements are non linear and the power supply of the circuit has a crucial role.while a given small signal model for the same circuit ignores possible simultaneous changes in the gain and power supply values.

Large Scale Model and Small Signal Model

The large-signal model is a model that is acceptably accurate over a large range on input signals. For transistors and diodes, this model is polynomial or exponential, which makes it difficult to work with. But if you restrict the signals to small variations, then over the range of those variations the response can be approximated very well as being linear, which is very easy to work with.

Basically the idea is that you have a circuit in which the transistor is biased at a particular quiescent point and the signal is causing to move about that point. So you break the total response into two pieces the DC bias, or operating, point and the variation due to the signal. Usually the bias point is large compared to the signal, so the circuit model used to calculate the DC operating point and the circuit that is used to model the changes about that point due to the signal. The first is called the large signal model and the second is called the small signal model.

References

- Ljiljana Trajković, "Nonlinear circuits", The Electrical Engineering Handbook (Ed: Wai-Kai Chen), pp. 79–81, Academic Press, 2005 ISBN 0-12-170960-4

- Distributed-element-model-494871: semanticscholar.org, Retrieved 25 July 2018

- Symbolic-circuit-analysis-691964: semanticscholar.org, Retrieved 19 March 2018

- Electronics, how-to-use-an-electrical-network-analyser-115545: streetdirectory.com, Retrieved 11 May 2018

- Kumar, Ankush; Vidhyadhiraja, N. S.; Kulkarni, G. U . (2017). "Current distribution in conducting nanowire networks". Journal of Applied Physics. 122: 045101. Bibcode:2017JAP...122d5101K. doi:10.1063/1.4985792

- What-is-the-difference-between-the-small-signal-and-large-signal-operation-of-a-BJT: quora.com, Retrieved 25 June 2018

- Explanation-of-small-signal-vs-large-signal-105987: allaboutcircuits.com, Retrieved 24 April 2018

5

Graphical Representation of Circuits

The graphical representation of an electrical circuit is called a circuit diagram. Such diagrams are useful for the design, construction, operation and maintenance of electronic and electrical equipment. This chapter closely examines the fundamental aspects of the graphical representation of circuits and includes topics such as circuit diagram, difference between schematics and circuit diagram and And-inverter graph.

Circuit Diagram

A circuit diagram (also known as an electrical diagram, elementary diagram, or electronic schematic) is a simplified conventional graphical representation of an electrical circuit. A pictorial circuit diagram uses simple images of components, while a schematic diagram shows the components of the circuit as simplified standard symbols; both types show the connections between the devices, including power and signal connections. Arrangement of the components interconnections on the diagram does not correspond to their physical locations in the finished device.

Unlike a block diagram or layout diagram, a circuit diagram shows the actual wire connections being used. The diagram does not show the physical arrangement of components. A drawing meant to depict what the physical arrangement of the wires and the components they connect is called "artwork" or "layout" or the "physical design."

Circuit diagrams are used for the design (circuit design), construction (such as PCB layout), and maintenance of electrical and electronic equipment.

Symbols

Circuit diagram symbols have differed from country to country and have changed over time, but are now to a large extent internationally standardized. Simple components often had symbols intended to represent some feature of the physical construction of the device. For example, the symbol for a resistor shown here dates back to the days when that component was made from a long piece of wire wrapped in such a manner as to not produce inductance, which would have made it a coil. These wire wound resistors are now used only in high-power applications, smaller resistors being cast from carbon composition (a mixture of carbon and filler) or fabricated as an insulating tube or chip coated with a metal film. The internationally standardized symbol for a resistor is therefore now simplified to an oblong, sometimes with the value in ohms written inside, instead of the zigzag symbol. A less common symbol is simply a series of peaks on one side of the line representing the conductor, rather than back-and-forth as shown here,

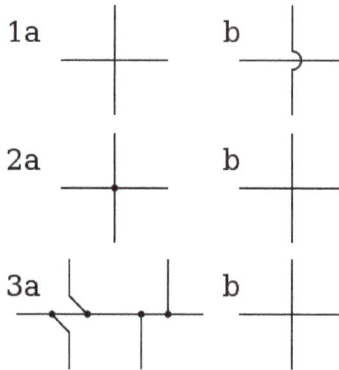

Schematic wire junctions: 1. Old style: (a) connection, (b) no connection.
2. One CAD style: (a) connection, (b) no connection.
3. Alternative CAD Style: (a) connection, (b) no connection.

The linkages between leads were once simple crossings of lines; one wire insulated from and "jumping over" another was indicated by it making a little semicircle over the other line. With the arrival of computerized drafting, a connection of two intersecting wires was shown by a crossing with a dot or "blob", and a crossover of insulated wires by a simple crossing without a dot. However, there was a danger of confusing these two representations if the dot was drawn too small or omitted. Modern practice is to avoid using the "crossover with dot" symbol, and to draw the wires meeting at two points instead of one. It is also common to use a hybrid style, showing connections as a cross with a dot while insulated crossings use the semicircle.

On a circuit diagram, the symbols for components are labelled with a descriptor or reference designator matching that on the list of parts. For example, C_1 is the first capacitor, L_1 is the first inductor, Q_1 is the first transistor, and R_1 is the first resistor (note that this is not written as a subscript, as in R_1, L_1,...). Often the value or type designation of the component is given on the diagram beside the part, but detailed specifications would go on the parts list.

Difference between Schematics and Circuit Diagram

A schematic circuit diagram represents the electrical system in the form of a diagram that shows the main features or relationships but not the details. In a schematic circuit diagram, the presentation of electrical components and wiring does not completely correspond to the physical arrangements in the real device. If you want to understand a schematic diagram, you are required to master basic knowledge of electricity and physics as well as internationally standardized symbols. Look at the parallel circuits below, you may find that the battery is represented as two short lines, the lights are represented as a circle with a cross inside, and the wiring is represented as a line. This kind of circuit diagram is mainly used by electrical engineers. The following is a schematic circuit diagram of a semiconductor electronic.

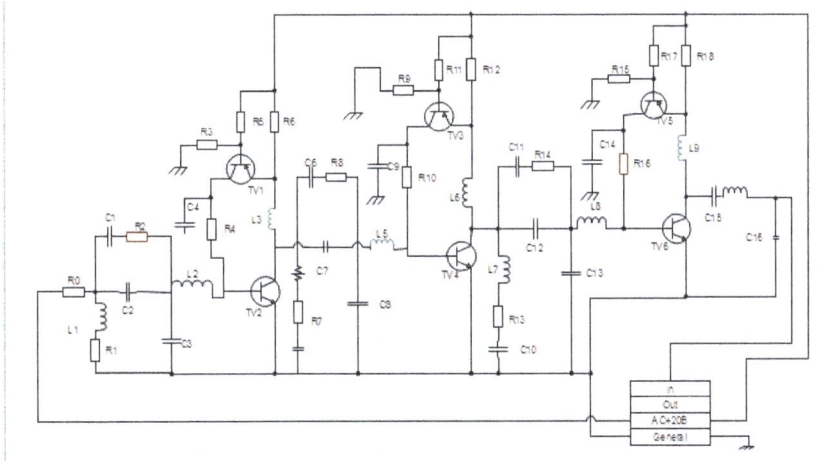

Schematics and circuit diagrams are both important engineering diagrams.

And-inverter Graph

An and-inverter graph (AIG) is a directed, acyclic graph that represents a structural implementation of the logical functionality of a circuit or network. An AIG consists of two-input nodes representing logical conjunction, terminal nodes labelled with variable names, and edges optionally containing markers indicating logical negation. This representation of a logic function is rarely structurally efficient for large circuits, but is an efficient representation for manipulation of Boolean functions. Typically, the abstract graph is represented as a data structure in software.

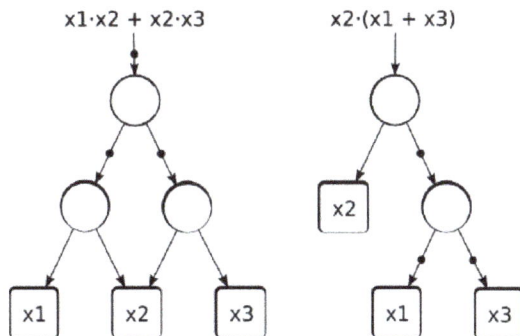

Two structurally different AIGs for the function $f(x_1, x_2, x_3) = x_2 * (x_1 + x_3)$

Conversion from the network of logic gates to AIGs is fast and scalable. It only requires that every gate be expressed in terms of AND gates and inverters. This conversion does not lead to unpredictable increase in memory use and runtime. This makes the AIG an efficient representation in comparison with either the binary decision diagram (BDD) or the "sum -of -product" (ΣoΠ) form, that is, the canonical form in Boolean algebra known as the disjunctive normal form (DNF). The BDD and DNF may also be viewed as circuits, but they involve formal constraints that deprive them of scalability. For example, ΣoΠs are circuits with at most two levels while BDDs are canonical, that is, they require that input variables be evaluated in the same order on all paths.

Circuits composed of simple gates, including AIGs, are an "ancient" research topic. The interest in AIGs started in the late 1950s and continued in the 1970s when various local transformations have been developed. These transformations were implemented in several logic synthesis and verification systems, such as Darringer and Smith, which reduce circuits to improve area and delay during synthesis, or to speed up formal equivalence checking. Several important techniques were discovered early at IBM, such as combining and reusing multi-input logic expressions and subexpressions, now known as structural hashing.

Recently there has been a renewed interest in AIGs as a functional representation for a variety of tasks in synthesis and verification. That is because representations popular in the 1990s (such as BDDs) have reached their limits of scalability in many of their applications. Another important development was the recent emergence of much more efficient boolean satisfiability (SAT) solvers. When coupled with AIGs as the circuit representation, they lead to remarkable speedups in solving a wide variety of boolean problems. AIGs found successful use in diverse EDA applications. A well-tuned combination of AIGs and boolean satisfiability made an impact on formal verification, including both model checking and equivalence checking.

Another recent work shows that efficient circuit compression techniques can be developed using AIGs. There is a growing understanding that logic and physical synthesis problems can be solved using AIGs simulation and Boolean satisfiability compute functional properties (such as symmetries) and node flexibilities (such as don't-cares, resubstitutions, and SPFDs).

This work shows that AIGs are a promising unifying representation, which can bridge logic synthesis, technology mapping, physical synthesis, and formal verification. This is, to a large extent, due to the simple and uniform structure of AIGs, which allow rewriting, simulation, mapping, placement, and verification to share the same data structure. In addition to combinational logic, AIGs have also been applied to sequential logic and sequential transformations. Specifically, the method of structural hashing was extended to work for AIGs with memory elements (such as D-type flip-flops with an initial state, which, in general, can be unknown) resulting in a data structure that is specifically tailored for applications related to retiming.

On-going research includes implementing a modern logic synthesis system completely based on AIGs. The prototype called ABC features an AIG package, several AIG-based synthesis and equivalence-checking techniques, as well as an experimental implementation of sequential synthesis. One such technique combines technology mapping and retiming in a single optimization step. It should be noted that these optimizations can be implemented using networks composed of arbitrary gates, but the use of AIGs makes them more scalable and easier to implement.

Permissions

All chapters in this book are published with permission under the Creative Commons Attribution Share Alike License or equivalent. Every chapter published in this book has been scrutinized by our experts. Their significance has been extensively debated. The topics covered herein carry significant information for a comprehensive understanding. They may even be implemented as practical applications or may be referred to as a beginning point for further studies.

We would like to thank the editorial team for lending their expertise to make the book truly unique. They have played a crucial role in the development of this book. Without their invaluable contributions this book wouldn't have been possible. They have made vital efforts to compile up to date information on the varied aspects of this subject to make this book a valuable addition to the collection of many professionals and students.

This book was conceptualized with the vision of imparting up-to-date and integrated information in this field. To ensure the same, a matchless editorial board was set up. Every individual on the board went through rigorous rounds of assessment to prove their worth. After which they invested a large part of their time researching and compiling the most relevant data for our readers.

The editorial board has been involved in producing this book since its inception. They have spent rigorous hours researching and exploring the diverse topics which have resulted in the successful publishing of this book. They have passed on their knowledge of decades through this book. To expedite this challenging task, the publisher supported the team at every step. A small team of assistant editors was also appointed to further simplify the editing procedure and attain best results for the readers.

Apart from the editorial board, the designing team has also invested a significant amount of their time in understanding the subject and creating the most relevant covers. They scrutinized every image to scout for the most suitable representation of the subject and create an appropriate cover for the book.

The publishing team has been an ardent support to the editorial, designing and production team. Their endless efforts to recruit the best for this project, has resulted in the accomplishment of this book. They are a veteran in the field of academics and their pool of knowledge is as vast as their experience in printing. Their expertise and guidance has proved useful at every step. Their uncompromising quality standards have made this book an exceptional effort. Their encouragement from time to time has been an inspiration for everyone.

The publisher and the editorial board hope that this book will prove to be a valuable piece of knowledge for students, practitioners and scholars across the globe.

Index

www.ingramcontent.com/pod-product-compliance
Lightning Source LLC
Chambersburg PA
CBHW062004190326
41458CB00009B/2961